The quest for
SUSTAINABLE AGRICULTURE
and land use

The quest for **SUSTAINABLE AGRICULTURE** *and land use*

Brian Roberts

UNSW
PRESS

Dedicated to my wife Margaret

Published in Australia by
UNIVERSITY OF NEW SOUTH WALES PRESS LTD
Sydney NSW 2052 Australia
Telephone (02) 398 8900
Facsimile (02) 398 3408

First published 1995

National Library of Australia
Cataloguing-in-Publication entry

Roberts, Brian (Brian Ross), 1933–
The quest for sustainable agriculture and land use.

Bibliography.
Includes index.
ISBN 0 86840 374 1.

1. Sustainable agriculture – Australia. 2. Land use, Rural – Australia.
I. Title.

630.994

Available in North America through:
ISBS Inc
5804 N.E. Hassalo Street
Portland Oregon 97213–3644
United States of America
Tel: (503) 287 3093
Fax: (503) 280 8832

Available in Singapore, Malaysia & Brunei through:
Publishers Marketing Services
Singapore 1232
Tel: (65) 256 5166
Fax: (65) 253 0008

Text formatted in 10.5/11pt Berkley Book
Printed by Southwood Press, Marrickville, NSW

6.10.96

CONTENTS

PREFACE

There is nothing new in the quest by humankind for permanent ways of using the land entrusted to them, but many planners regret the day that the World Council of Churches proposed the term 'sustainable' to describe the objectives of global society.

The challenge in writing this book was to identify which players, policies and procedures could contribute to sustainable societies globally and, through international connectedness, to sustainable rural production in Australia. As each building block in the sustainability debate was identified, its position in the grand scheme of things was considered. It appeared that the sequence in which facts from the volume of published material are recorded is less important than their interconnectedness. Once the researcher overcomes what Odum called 'the struggle for perspective', a pattern and logical structure appears. But the uncertainties of absolute timeframes and levels of permanence remain. This is inherent in the nature of the study material, so to cater for both the cornucopian (God and the technician will see us right) and the deep ecologist (pristine interconnectedness is the goal), the researcher offers a range of future paths. This application of situation ethics avoids the embarrassment of having to defend a single position in a debate where no absolute values can survive on their own. Such flexible values can also be seen as vague, even contradictory, in a world seeking clear guiding lights in a sea of uncertainty.

The alternative is to select a particular position which requires moral judgements on what people need and thus on what is right. Such a position on sustainability assumes predetermined levels of complex attributes of the human condition, notably quality of life, as distinct from material living standards. For while we believe that we know what has to be done to bring order and peace to our world, we feel powerless to act, at least on the global scene. Charles Birch, perhaps the greatest of Australian thinkers on these matters, suggests that the age of absolute national sovereignty is over but the age of international order is not yet born.

Sustainable agriculture is usually dealt with as if it is a matter of selecting appropriate technologies, but it soon becomes clear on examining the externalities which impinge on the subject, that it is primarily a matter of fundamental value choices. These are choices of priorities in space and time for we are dealing with interdependent trading nations attempting to pursue permanent land-based production.

For these reasons this book consciously avoids a detailed analysis of the mechanistic and chemical approaches to sustainable agriculture. Instead it asks many questions, encouraging the reader to consider the should/ought decisions rather than the can/do abilities of modern humankind. The need for action in Australia was covered in *The Birth of Land Care*; the symptoms, processes and corrective actions are the thrust of the *Land Care Manual*. The present study attempts to paint the big picture, setting the scene nationally and internationally in terms of conditions and actions necessary to make at least an approach to biological permanence in agriculture, as championed in *Ground Rules: Perspectives on Land Stewardship*.

The author is indebted to Rex Parry who put forward the challenge to undertake this work; to Marie Schulz, Secretary to the Land Use Study Centre, for her word processing and printing of several drafts of a difficult manuscript; to the librarians of the University of Southern Queensland for literature searches and to the University of New South Wales Press for editing and publishing.

B R ROBERTS
Land Use Study Centre
University of Southern Queensland

Chapter 1

LAND AND SOCIETY: GLOBAL RELATIONS

This chapter sketches the global situation as an introduction to the concept of planned resource use which is dealt with in Chapter 2. Here we deal with the impact of what is commonly termed 'progress' and we reconsider humankind's priorities in modern development. The stark comparison of the 'haves' and 'have nots' of our global population is used as a basis for considering our so-called 'common future' which emphasises that we're all in this global situation together. The need to seek international cooperation in problem-solving is emphasised as a fundamental element of securing our global future. The chapter concludes by examining the role of multinational companies and their worldwide effects on development and on the maintenance of environmental quality.

1.1 Global priorities — progress and impact

For some 40 years the threat of nuclear war and its consequent nuclear winter have dominated the human prospect. This has changed in recent years, thank God (or as some say, thank Gorbachev). McHarg, author of the classic *Design With Nature* (1969), states that nations committed to expurging each other now assert that their primary purpose is to nurture the earth. This has resulted from an increasing awareness of the human population exceeding 5 billion, species extinction, pollution and a realisation of the effects of climatic change resulting from the greenhouse effect. Looking back, we may credit Osborne with his *Our Crowded Planet* (1963) and Rachael Carson with her *Silent Spring* (1962) for initiating the western environmental revolution. McHarg (1992) today emphasises that this revolution was not led by 'the establishment'. The US departments, institutes and scientific societies were mute. The major religions did not participate. It was the young, the sensitive and the radical who challenged the status quo and shocked officialdom out of complacency. Today the questions must be asked: Are human beings like pervasive global pathogens, agents of planetary disease? Can a human become an enzyme, a catalyst, in maintaining the earth's crucial processes? Can humans exploit their unique capability of consciousness and become earth's physicians, earth's healers?

What of Australia's priorities? Is this nation following the mistakes of other western societies with the usual timelag? Or is the Clever Country astute enough to heed the early warning signals of ecosystem breakdown, and channel its resources into preserving its productive landscapes? Is Cribb's (1990) assertion that we can become 'clean, green and clever' well founded or does the dirty, brown and stupid alternative still linger if we look at the lessons of history?

The effects of humanity on the earth depend largely on how societies are organised and controlled. In essence the integrity of the earth cannot be restored and sustained without attention to the pillars of our civilisation:

- values and beliefs
- cultural and artistic norms
- knowledge generation and transfer
- research and development
- business and financial institutions
- social and other service systems
- legal and justice systems
- civic and political institutions.

Calvin De Witt (1987) has suggested that because our institutions have developed largely in the absence of global environmental awareness, humanity has not adapted society to the needs of the globe. Stephen Boyden (1991) of the Australian National University suggests that the contemporary 'cultural soup' of modern western society is, indeed, dangerously deficient in a crucially important ingredient. This missing ingredient is an understanding by the majority of the population of:

- nature
- the sensitivity and interdependence of living systems
- the place of humans in nature
- relations between cultural and natural processes.

Figure 1.1 shows relationships between humans, culture and environment.

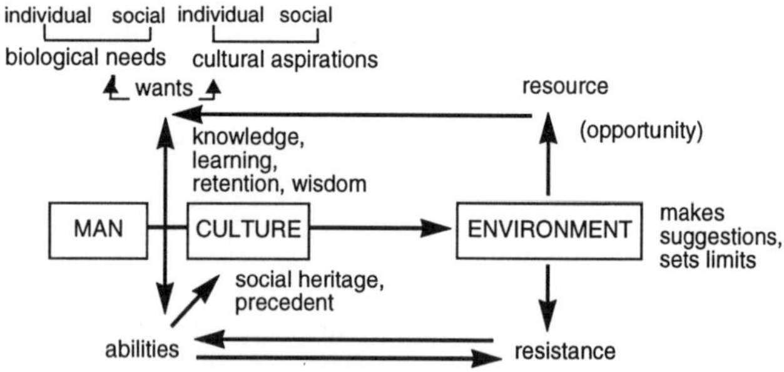

Figure 1.1 The relationships between culture and environment
O'Riordan, 1971

Boyden maintains that correcting this deficiency is essential for attaining an ecologically sustainable society which is at the same time satisfying in terms of quality of life. Such attainment will require a *biosophical* approach to priority-setting and decision-making throughout society.

Global society is affecting our finite planet in three different ways:

1 By the numbers of people the earth is expected to sustain.
2 By the levels of consumption of ever more materialistic societies, which harvest renewable resources beyond their sustainable harvest rate.
3 By pollutants or byproducts, including an increasing range of synthetic products which nature cannot decompose.

The change from present material consumptive varieties requires a transitional stage in which societies change the values, ask different questions and change their behaviour. Brown (1990) sums up these contrasting values in Table 1.1.

Table 1.1 Value contrasts in pretransitional and transitional societies

	Pretransitional	*Transitional*
Organising principle of economy	Materialism	Sustainability
Nature of product	Planned obsolescence; throwaway mentality	Durability, high quality engineering and design
Relationship to nature	Domination of nature	Harmony with nature
Scale	Bigger is better	Smaller has a place too
Speed	Let's go faster	Where are we going?
Specialisation	Key to efficiency	Increases dependency, leads to boring work
Determinants of status	Material possessions	Personal development, social contribution
Childbearing	Taken for granted, virtually automatic	Optional; to be given serious thought

Source: Brown, 1990

This overview focuses on land as a basic resource and on how land, notably in Australia, is used for agricultural and pastoral purposes. When land use alternatives are evaluated, agriculture must be compared with forestry, tourism and mining as legitimate competing uses of rural land.

1.2 The haves and have nots of global agriculture

Recent population trends indicate that each year the world must feed an additional 93 million people despite the decline in cropland available to about 0.1 ha per person. Cultivated land makes up about 11% of the earth's land surface. Efforts are being made to expand our global food production area by deforestation for crop production in high rainfall zones, and by the expansion of *marginal* cropland in low rainfall zones. The dilemma is that while soil degradation (reduction of productive capacity) already occurs at a moderate rate in many of the world's best cropping areas, attempts to expand the frontiers of grain production often lead to a higher rate of degradation in the new lands, whether wet or dry.

A further dilemma facing today's agricultural producers is that the application of large amounts of agricultural chemicals (fertilisers, pesticides and herbicides) means that cropping systems depend on higher and higher inputs of chemicals and energy. These matters will be dealt with in detail in later chapters. At this point it is sufficient to appreciate the general spiral of dependence on expensive *synthetic* (unnatural or man-made) compounds which, having served their short-term purpose, become long-term pollutants in the global ecosystem. The challenge to global land use planners is thus how to allocate each portion of the world's 13B ha of total global land surface area to optional land uses that will maintain the productive capacity and the stability of that land for the benefit of future generations.

Even a cursory examination of standards of living in the world's hemispheres indicates that the North's environmental problems are primarily due to consumption and pollution, while the South's problems mostly concern population growth and food supply. While the North consumes a disproportionate amount of global resources, it is also responsible for air pollution which threatens global climate to the detriment of all societies. The South on the other hand is having a major effect on global food production potential through erosion caused by deforestation and the extension of agriculture into steep fragile monsoon lands.

Any treatise on land use and agriculture must take both the broad view geographically, and the long view time-wise if it is to consider alternative land uses and production systems in their full context. Indeed it is the failure to recognise the exploitative nature of many recommended agricultural systems in the past which has led to social collapse and poverty. Accounts such as Carter and Dale's (1974) *Topsoil and Civilisation* and Jacks and Whyte's (1939) *Rape of the Earth* have long been recognised as reliable chronicles of human folly down through the history of land management. This book attempts to catalogue what we have learned from this hard lesson of the past and how, with present knowledge, we can develop land use systems that use each type of landscape according to its inherent capacity and maintain the soil resource in a permanently 'healthy' condition.

The World Resources Institute (1985) has given a useful overview of the generalised trends in the condition of the world's productive rural land. Fifteen years ago two-thirds of the population of the developing countries already lived in countries rated 'land scarce' or 'extremely land scarce', i.e. where more than 70% of potential cropland is already cultivated. The best estimates indicate that by the year 2000 countries with industrial market economies will have about 0.5 ha of cropland available per person. This area drops to 0.25 ha in centrally planned economies (e.g. China) and to just over 0.1 ha in the developing countries of Asia and Africa. In Asia the major form of land degradation is water erosion, because more and more steep country is being deforested and cultivated due to population pressure. The same pressure had led to replacing the earlier slash-and-burn system of land rotation with continuous annual cropping and the resultant depletion of the soil. In Africa the process of 'desertification' is spreading by the practice of cultivating droughty shallow marginal soils annually, and by overstocking sparse fragile semi-arid grazingland. Both processes cause severe erosion and soil deterioration. The importance of the length of the soil restoration periods is reflected in Figure 1.2.

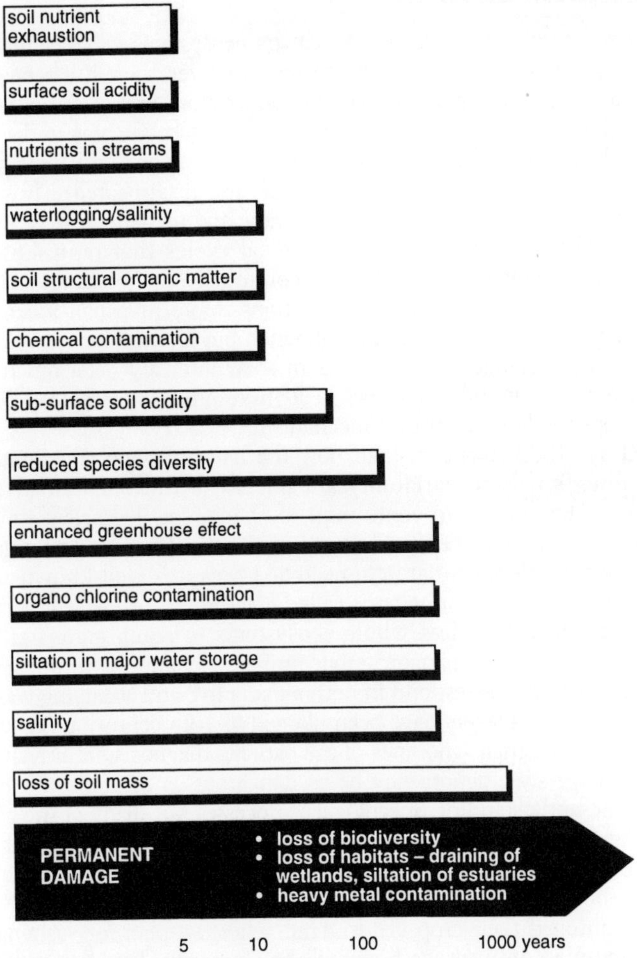

Figure 1.2 Timescales in land degradation
WGSA, 1991

After erosion, the problem of salinity, especially on irrigation schemes, is the major form of land degradation caused by improper land use. The World Resources Institute's latest estimates are that irrigation land is being lost to salinity at much the same rate at which new irrigation schemes are being developed worldwide. The gravity of the land degradation situation is reflected by the world's two largest nations: China has lost 11% of its cropland to erosion, desertification and to non-agricultural uses since 1957, while 25% of India's total land area is affected by erosion (13M ha by wind and 74M ha by water).

1.3 Sustainability and change in the natural environment

There is a tendency to consider the earth's ecosystems as constant and unchanging, when left to natural processes. However, there has been continuous, albeit natural, modification and adaptation, first to temperature changes, then to moisture changes. Thus ice ages have come and gone, the earth has cooled then warmed, and the ecosystems and their biota have adapted and evolved, or drawn back and disappeared. Thus not only the more obvious forms of the landscape and their dependent living organisms change, but the biogeochemical cycles that represent the lifeblood of our global ecosystems also continually change as an adaptation to physical environmental modifications. More than half a century ago the Yugoslav mathematician Milankovitch put forward his remarkable hypothesis that changes in the effects of solar intensity determined the rate at which snow and ice sheets melt in summer, and that all other changes depended on this trigger mechanism.

McElroy (1985) has explained how the nature, origin and limits of the biosphere's natural variability are related to this solar insulation hypothesis. This long-term continuum of change has led scientists like James Lovelock to view the world as one interlocking system, one organic whole in which all things are connected. Lovelock's well-known *Gaia* principle is based partly on what is called *homeostatis*, or the tendency of single organisms as well as whole ecosystems to reach a constant or steady state. Thus while macro-environmental change continues, the earth and its organisms respond in both an adaptive and steadying way.

Today's land managers have been likened to 18th century physicians, for while we share their ignorance of the patient's diagnosis, intuitively we sense that large-scale action may be as dangerous as inaction. We have gained some understanding of ecological processes but many of the complexities and side-effects of our actions are poorly understood. We have learned by trial and error, like the Romans who eliminated malaria by draining the swamps or the Greek farmers who did not know why legumes improved their crop yields. Thus when we plan for sustainable societies founded on what we hope will be permanent land use systems, we have no option but to use our incomplete knowledge of the chain reactions occurring in our changing biosphere — that narrow band of life on the circumference of our planet.

1.4 Our common future

This phrase is the title of what is perhaps the most quoted of global reports on the destiny of humankind, i.e. the Report of the World Commission on Environment and Development (the *Brundtland Report* of 1988). It was this document that sparked first the common use of the term '*sustainable development*', then the widespread attempts of governments worldwide to at least initiate plans for the sustainable development of their countries. The World Commission defined as sustainable '*that development which meets the needs of the present without compromising the ability of future generations to meet their own needs*'.

Predictably, such a simple definition of a complex global issue has generated a running debate about all the 'ifs and buts' that arise in defining the concepts and the values associated with their legitimate alternative interpretations. Since World War II international development has aimed to improve the standard of living of all peoples, as measured by income, gross national product, health, life expectancy and infant mortality rates. The western view of the world usually assumes that development is synonymous with economic growth. Thus development has come to be measured as growth and wealth in monetary terms. Too often a decline in environmental standards and quality of life was accepted as part of the price to be paid for so-called *progress* in the guise of industrial development.

In recent years sustainable development as an attainable objective of planning has overtaken economic growth at any cost as the prime goal of countries concerned with the future quality of their citizens' lives. One of the most telling demonstrations of this fundamental change in direction was the alliance of non-government organisations (NGOs), foreign aid agencies and conservation groups at the Rio Global Forum (Roberts, 1992d).

Commins (1989) suggests that the ideal of *sustainable development* is a concept that combines two conflicting visions of social change. It is argued that sustainable development can be attained without destroying the environment while improving the livelihood of low-income people. This ideal originates from environmental concerns rather than financial considerations and is largely based on the conviction that present economic practices, especially as applied in modern agriculture, cannot preserve the earth's productive potential for future citizens of the planet.

Both capitalist and socialist systems are based on the quest for economic growth, which in turn is recognised as the mark of progress — the measure by which a nation's wellbeing has been gauged. Since the 1960s, however, people have come to realise that planning based only on economic realities neglects ecological issues. Thus policies aimed at satisfying short-term economic and social needs have conflicted with the need to account for what has become known as '*intergenerational equity*'. The destruction of the resource base is becoming an unacceptable trade-off for improved living conditions. Figure 1.3 presents a framework for considering the major problems affecting the global human condition.

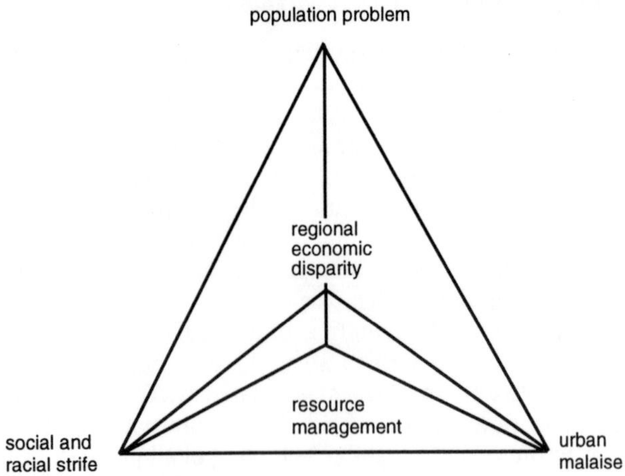

Figure 1.3 A framework for resource problem-solving
O'Riordan, 1971

This crucial change in social values came about as we became aware that humankind was living beyond its means ecologically, and that the interim benefits of consumerism were gained more and more at the expense of a productive and stable earth. This shift in values was encapsulated by the World Commission (Brundtland, 1988) which was founded in 1983 specifically to examine the interrelatedness of environmental and developmental issues. These terms of reference were in marked contrast to the Stockholm Report (UN, 1972) which focused solely on environmental trends, and the Brandt Commission of the same period, which produced two seminal reports on global poverty and development entitled 'North-South' and 'Common Crisis'. In its introduction, the 1978 World Commission report noted:

> We have in the past, been concerned about the impact of economic growth upon the environment. We are now forced to concern ourselves with the impacts of ecological stress — degradation of soils, water regimes, atmosphere and forest — upon our economic prospects.

Perspectives on the need to develop social and production systems that will maintain *environmental quality* (productivity and stability) are now emerging in several current social changes. One such change is the conservation lobby's acknowledgement that developing countries destroy their environment mostly because they are driven by economic forces beyond their control. Another change is that environmental groups and community aid organisations work together to gain recognition of the fact that environmental protection is no longer a luxury to be pursued only by the wealthy, but is now also a survival mechanism required by the poor and, in time, by the total global family.

1.5 World Bank and the multinationals

Development has largely depended on financial resources from international lending institutions. The World Bank and others such as the Inter-American Development Bank have mostly controlled the location, direction and size of development projects in the Third World. The loan policies of these lending institutions have been challenged in the light of accumulating evidence of environmental damage caused by World Bank-funded projects and allied programs. Since the late 1970s serious doubts have arisen about the funding priorities and the loan practices employed by major institutions which make scarce capital available to poor countries. The sustainability of many projects is questioned on three grounds: (1) on the basis of who benefits from major projects such as dams, factories or communications; (2) on the disastrous environmental effects caused by the inappropriate exploitation of natural resources such as forests, fish stocks and grazing land; and (3) on the pollution of water, air and soil caused by the practices and processes of development projects in densely populated regions.

Salter (1989) has argued that the poor must put survival first, that they are not in a position to consider long-term or even medium-term consequences of their fight for survival. Thus many multinational companies find it convenient to operate both their manufacturing and sales divisions in developing countries that cannot afford to enforce environmental control regulations. This global movement of environmentally damaging practices — notably in the chemical industry, but also in the mining and energy sectors — into Asia, Africa and South America is almost entirely financed by developed countries in the West. The transfer of wealth that such activities generates only exaggerates the inequalities between the world's rich and poor. Figure 1.4 reflects the potential biological productivity of the developing world, indicating that in the three major developing regions poverty prevails despite high potential.

Parallel with this exploitation of the environment of poorer nations is the debt crisis which has put these nations under the financial control of the wealthy industrialised nations. Presently the Third World is faced with a combined debt of over 1 trillion dollars ($1,000B), a debt level that consumes virtually the entire foreign earnings of many small nations simply to repay interest on foreign loans. In this situation, such nations understandably give no priority to investing in pollution control or other sustainability elements in their production systems.

In the great majority of cases the rural poor in the Third World have only two options — remain as landless labourers suffering chronic poverty and malnutrition, or take the opportunity to occupy newly deforested or mountainous land. With most of the Third World's good agricultural land concentrated in the hands of an elite class of landlords, the redistribution of land becomes critical, both to social justice and to the sustainability of societies.

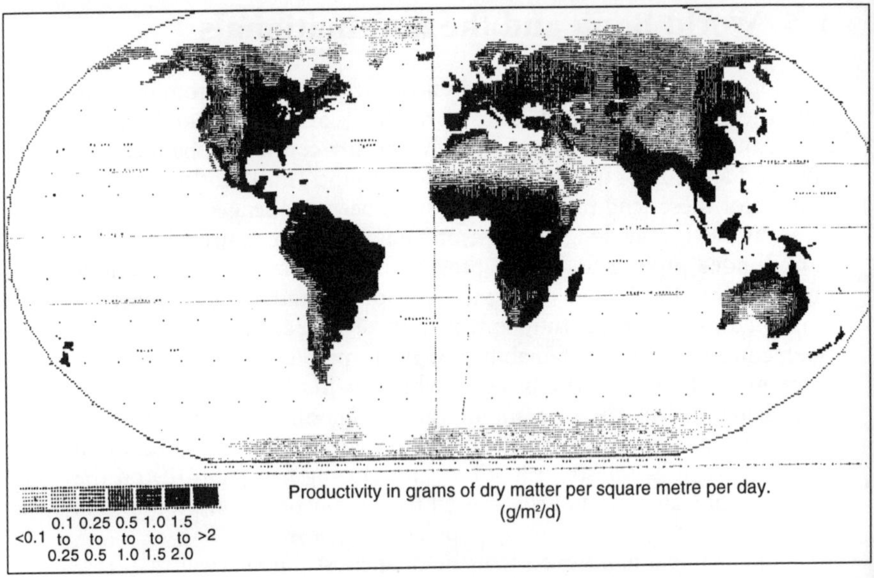

Figure 1.4 Inherent productivity of the world's land
White, Mottershead and Harrison, 1984

Chapter 2

PEOPLE, RESOURCES AND THE CONCEPT OF SUSTAINABILITY

Chapter 1 gave a brief overview of the direction in which our global societies are heading. This chapter concentrates on our dependence on our land resources and the need for their intelligent stewardship. Here we define what is meant by sustainable societies and how their quality of life largely depends on planned and sensitive resource management. The needs of both present and future populations are considered, together with proper accounting of the impacts of optional land uses. We reconsider the economic effects of neglecting the condition and conservation of our natural resources, and stress the need for special attention to fragile ecosystems such as occur in semi-arid zones. The chapter reiterates the need for understanding the responsibilities and susceptibilities of the ecosystems on which our future quality of life depends.

2.1 First the sustainable society — what is it?

To be sustainable is to be capable of indefinite existence. On a finite planet this means replacing a *growth economy* with a *steady state economy*, or in Boulding's (1971) words replacing a *cowboy economy* (wasteful, extravagant) with a *space-ship economy* (frugal, re-use, clean). Since the negative connotations of 'steady state' were interpreted as 'stagnant' or 'no-growth' by the developing countries, the World Council of Churches (1974) coined the phrase *sustainable society* to emphasise the conservation of life-support systems.

Birch and Cobb (1981) suggest that sustainability in the real world is a relative term, like justice or equity, and they suggest that we will do well if we can define social systems that will last even a few hundred years. Sustainable does not imply static, but rather changeable, dynamic, adaptable, but not self-consuming. So the sustainable society will have the following characteristics:

1 The *population* will be within the land's carrying capacity as modified by economic structures.
2 The demand for *food, water and other renewable resources* will be within the capacity of the land to supply such demand.
3 The *emission of pollutants* will be within the capacity of the ecosystem to absorb them.
4 The rate of *consumption of non-renewable resources* such as energy and minerals will allow replacement technology to develop alternative sources.
5 *Manufacturing* will be based on long-lasting, recyclable goods.
6 *Social systems* will enhance stability through participation and equity.
7 *Quality of life* will depend on environmental quality and services, rather than material goods and acquisition. (Birch and Cobb, 1981)

The more specific attributes of sustainable agriculture must be defined within the broader goals of such a society. Figure 2.1 represents how various social goals overlap and how this tendency requires simultaneous consideration in national development planning.

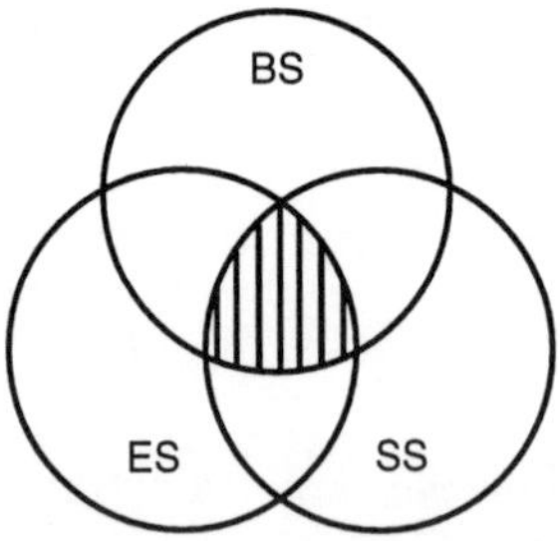

Figure 2.1 Interrelations between three major systems with common goals. BS: resource system, ES: economic system, SS: social system.
Source: Barbier, 1991

2.2 Sustainable populations

The term 'sustainable' implies maintaining or permanently supporting a particular situation or system. As such, it must make assumptions about the level of demand for the earth's products in relation to what our finite globe can supply. The relation between human population and the food production capacity of the earth is a case in point.

Given equitable and efficient distribution, the earth can probably supply the nutrient requirements of a total global population considerably larger than the 5.3B humans presently inhabiting it. However, population increase must be recognised as the ultimate fundamental threat to the global environment, since consumption per capita, pollution control and the renewal of resources (and our control of them) will eventually determine the earth's human carrying capacity. Many analysts have asked whether there is a simple relationship between poverty and over-population:

which causes which? The need for farm labour and for care of the elderly have repeatedly been advanced as reasons why Third World societies favour large families. In the West, however, the trend toward zero population growth (ZPG) did not commence until an acceptably high standard of living had been achieved. Until the widespread inequalities of opportunity have been overcome on a global scale, it is unlikely that a satisfactory reduction in overall population growth will be achieved — nor is it likely that the Third World pressures on natural resources and the environment in general will diminish until such inequalities are redressed.

Although it has become fashionable and environmentally correct to favour sustainability as a generalised goal for society, the meaning and implication of a sustainable society remain vague and contentious. Tisdell (1990) suggests that the implicit goals of different proponents of sustainability indicate that they aim to sustain different things and different situations. Thus equality, economic welfare, equity for future generations, and ecological stability and biodiversity emerge as some of the aims of sustainable development. Economists argue that the maintenance of natural environments, which often appears as the major motive for sustainable development, is not likely to succeed without intervention in the marketplace. If Rawl's theory of justice based on intergenerational equity is accepted as the justification for sustainable development, then such equity should also apply to different groups or nations. Distinguishing 'needs' from 'wants' is a crucial equity issue that must be settled before determining the size of the global population to be sustained.

In simple terms, Sears' (1959) formula relates population to standard of living in the following relationship:

$$S = \frac{R \times T}{P}$$

where S is standard of living, R is resources available, T is technology available and P is population. While simplistic, Sears' (1959) relationship is soundly based: irrespective of the levels of natural resources and exploitation technology, the greater the population the lower the individual standard of living, at least in material terms. Its implication for sustainability is that the quality of life that can be maintained depends equally on the three determinants of that quality.

In my view, the emphasis on sharing, opportunity and choices too often overshadows the concepts of reward for effort, just deserts and personal gain from initiative, foresight and ability. The emphasis on equity for humans also virtually eliminates the ecologically reasonable value-system of those such as Nash (1991) who claim that living organisms other than the human species also have standing, and thus also have a claim to equity or at least existence.

Besides the question of what is equitable, the question of what levels of population and resource use are sustainable must be answered before developmental targets can be deemed as achievable in a biophysical sense. The American economist Hermann Daly has offered a mixture of steady-state economics and Christian pragmatism as the cure for planetary degradation. In the same way, Dennis Meadows and his colleagues, in their now classical benchmark study of the early 1970s, *Limits to Growth*, warned that industrial production, population and consumption must be brought into equilibrium with pollution and food production if a lasting global society was to be achieved. In the more recent evaluation of global trends, *Beyond the Limits*, Meadows et al. (1992) insist that rather than being too pessimistic, their predictions made 25 years ago in fact underestimated the severity of human impact on global ecological sustainability.

Theoretically a sustainable society will eventually have to do without non-renewable resources such as fossil fuels and heavy metals. While this view may seem extreme and unreasonable in the medium term, commentators such as Georgescue-Roegen (1980) believe that the human race has its best chance of succeeding as a permanent global species if it learns to maintain its societies without dependence on dwindling non-renewables. This stance argues that population must be limited to a level that can be sustained by renewable resources only. This brings us to considering resources as the basis for humankind's future.

2.3 Resources as opportunities and options

There is no universally satisfactory definition of a resource. The closest we may come is summarised by O'Riordan (1971) as *an attribute of the environment, appraised by man to be of value over time, within constraints imposed by his social, political, economic and institutional framework*. Thus as a consequence of the social and technological cultures of each group, gold was not a resource to the Aborigines and kangaroos are not a resource to Australian farmers. For this reason Zimmerman (1951) suggests that rather than regarding a resource as a tangible object, it should be viewed as a functional relationship between humans' wants, abilities and environmental capability. He points out that our technological exploitation is constrained by 'ecological resistance' which ultimately limits our activities to an equilibrium with nature.

Before compiling the building blocks of a sustainable system, resources must be categorised. Highsmith (1969) gives three main resource types:

1 *Flow resources*, which can be sustained, depleted or increased by management, e.g. soils, fauna, flora, water.
2 *Stock resources*, which are either (a) depletable but reusable (e.g. metals) or (b) totally exhaustible (e.g. fossil fuels).
3 *Continuous resources*, which are either (a) infinitely available irrespective of human action (e.g. solar and tidal energy) or (b) infinite but influenced by management (e.g. space, air and landscape).

When these categories are compared to classical divisions of resources into land, labour and capital, the concept of sustainability becomes easier to integrate with modern planning than it was with the previous utilitarian and human-centred value system.

O'Riordan (1971), in his useful overview of the origins and development of the Western conservation philosophy, points out that our appreciation of nature arose from a scarcity of resources rather than environmental sensitivity. Thus the concept of sustainable development developed as a reaction against unthinking exploitation, but not before a long period of polarisation of producers and conservationists had passed. Today the sustainable development lobby represents the common ground in the middle of the continuum of views from exploitation to preservation. Figure 2.2 displays four major areas of decision-making in this continuum of conservation.

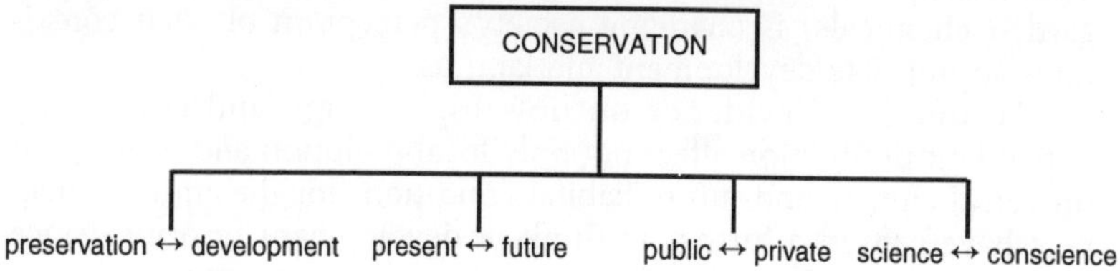

Figure 2.2 The four basic dichotomies in conservation
O'Riordan, 1971

2.4 Resources and quality of life

Sustainable land use must be approached from two viewpoints — one considers land resources as wealth-generating production factors; the other considers the contribution of landscapes to our quality of life through environmental aesthetics and ecological diversity.

Previous generations regarded land very much as the source of potential riches through saleable earth products. Recent decades have witnessed a great increase in the demand, not for raw materials from the land, but for services and amenities. At the same time value-adding has overtaken primary production (agriculture and mining) as the major component of commodity price. These changes in resource evaluation have been accompanied by an emerging environmental awareness. This appreciation of what landscapes can contribute to our quality of life has been central to the new values now accorded to natural ecosystems. Thus in current land use planning, rainforest is more than a timber resource, wild rivers are more than a hydro-electricity source and mangroves are no longer regarded as useless swamp areas fit only for drainage and development. Within this enlightened planning framework, land resource planning is fraught with a range of value judgements which reflect a trade-off between land resources as wealth-generating raw materials and landscapes as increasingly important contributions to quality of life.

Economists have long recognised the tendency of societies to place higher values on luxury items as affluence increases.

Environmental quality and the aesthetic appreciation of nature is one such item, and it features in the list of criteria of a sustainable society — thus the catchcry of modern conservation groups that they are not blindly opposed to development but are opposed to blind development. There is a growing demand for sustainable development to supply society's material needs (food, energy, materials) while also safeguarding the human habitat. Western societies such as Australia's are not only well informed, through their home-delivered media, of the deterioration of global ecosystems but have already initiated moves toward a society that lives more gently on the earth. A reduction in consumerism combined with an increase in recycling and a sensitivity toward 'unfriendly' products (detergents, propellants, garden chemicals) is changing society's perception of what constitutes appropriate development and land use.

Accumulated evidence on how high-energy and chemically dependent production affect not only local pollution and health, but universal climate and future habitat conditions for the entire human race have caused a further re-think of development priorities since the early 1980s. The relations between *private benefit* and *public cost* have developed into a new form of environmental accounting. Through the process of full ecological accounting of environmental costs, social costs and the 'public good' are being factored into national budgets, so that degradation costs are no longer regarded as 'free goods'.

One of the conflicts in assigning resources to various uses concerns the value judgements related to resources for wealth generation (e.g. agriculture) compared to resources for aesthetic or altruistic purposes. Griffiths (1991), in tracing the conflicts in conservation movements, suggests that the first conservationists had been champions of the 'wise use' of resources. This was followed by a movement to preserve 'spiritually uplifting' and 'aesthetically pleasing' places, which he suggests shifted the emphasis to non-human values. Over time, the quality of life came to include not just immediate human comforts, but the preservation of biodiversity, the protection of gene pools, the integrity of ecosystems and the rights of plants and animals. As these new priorities evolved, they became reflected in the types of national parks and reserves that were being set aside in Australia in the late 1800s. More recently, areas of biological uniqueness have been included, irrespective of their scenic value (e.g. the Mulga and Mitchell grass ecosystems of Queensland).

Resource allocation for agriculture and pastoralism has thus faced increasing competition from biological and aesthetic alternatives including the intangible spiritual values of wilderness. These values were portrayed in their modern context by Nash (1982) in his *Wilderness and the American Mind*. Nash's approach, which considers wilderness as a place where humans are intruders, redefines their rights and political justification for claiming resources for agriculture, mining or forestry. From this change in attitudes, the direct conflict in agricultural land rights has arisen. Where it was once assumed that wild nature must be changed or even eliminated if quality of life and civilised living were to flourish, latterday values exalt the wilderness environment to the point where it is now defined by the absence of human intervention. Brown (1982) proposed that wilderness be defined as 'a region of original Earth where one stands with the senses entirely steeped in nature and free of the distractions of modern technology'.

This change in the view of wild land from 'worthless' to 'wonderful' is likely to continue its major influence on the agricultural industry's access to land in Australia. The confusion of values is compounded when Australia's various racial cultures converge in the resource/wilderness debate. This was well demonstrated in the Tasmanian Gordon-below-Franklin dams conflict when conservationists dominated Aboriginal claims to the area. Whereas museums formerly placed Aborigines in nature as 'primitive savages', they now tend to reverently reclassify them as 'ecological beings' (Griffiths, 1991). The reasons why the present Tasmanian wilderness is uninhabited are best forgotten for political reasons.

In considering resources in the quality of life debate, the stand taken by the Victorian high plains cattlemen against the Alpine National Park (declared in 1989) is an important case study in land rights and social perceptions of appropriate land use. The 'mountain men' represent rugged symbols of a pioneering community's sense of attachment to their land, akin to that of the Aborigines displayed in their way of life and professed understanding of their habitat and its requirements.

The Snowy Mountains conflict between pastoralism and parks raises a third dimension to land resource allocation — that of 'cultural landscapes', or areas reflecting historic land use (stockmen's huts, gold mining sites, timber camps etc.). Land use planners are faced with the complex task of weighing these values against those of wilderness and future wealth generation. Matching the rational and scientific to the nostalgic and romantic is an evaluator's nightmare.

Are opportunities to commune with nature necessary to what constitutes a sustainable society? The idea of nourishing the human spirit in a wilderness atmosphere is regarded by many as an essential ingredient of a balanced and unstressed lifestyle, but in an ever more crowded world (93M more people each year) such spiritual enhancement may be a luxury that few can aspire to. The preservation of wilderness in the form of national parks is a worthy international goal and is theoretically aimed at 5% of the land area of nation states. As described elsewhere, the wilderness debate does not draw on absolute values or quantities, but rather on the general desirability of recognising the benefits of wilderness as a legitimate land use alternative. There are no single best answers to questions about how much rainforest should be reserved, nor is it likely that today's best guess would be acceptable to tomorrow's generation.

The psychological benefits of wilderness relate to self-sufficiency as well as spirituality, as epitomised by the late Olegas Truchanas, father of the green consciousness in Tasmania (Angus, 1975). If a well developed sensitivity toward other living beings is the ultimate measure of spiritual development and thus of quality of life, then wilderness can contribute to the development of this element of a sustainable society. The human sciences have yet to evaluate the tangible benefits of individuals being alone for significant periods in wild surroundings. Australia, as one of the developed countries best placed to offer wilderness experiences to a small population probably has a greater opportunity to embed this element of quality into its future society's lifestyle.

2.5 Sustainable resource management — for whom?

Land use policies are based on complex relationships between socio-economic needs, physical resources and the biological requirements of the landscape's renewable elements. O'Riordan (1971) summarises the complexities of a sustainable land use plan as follows:

> The concern is to devise a mechanism by which we can evaluate the benefits (desirable streams of goods and services) and costs (a combination of capital expenditure and undesirable side effects) of alternative resource combinations and management techniques in relation to the achievement of desired goals.

The land use plan, whether local, regional or national, must determine how each option benefits and costs the different social groups affected, both now and later. In practice, planners give most weight to the needs of local communities in the short and medium term. Vision and statesmanship, which encompass other societies far away and later on, are rarely rewarded by the current democratic system in western consumer societies.

In blunt terms, the history of the developed countries is a catalogue of short-sighted programs driven by greed, ecological misunderstanding and political expediency. Too often the quest for growth, jobs and industrialisation caused a cycle of unsustainable reactions which almost invariably led to ecological breakdown and environmental deterioration. How then can the developmental options be compared in terms of costs, benefits, practical feasibility and financial efficiency? One popular approach has been the well-known '*cost-benefit analysis*' which originated in the United States in the mid-1950s. As computerised manipulation of large volumes of data (particularly multivariate data) became easier, so the technique of systems analysis came into vogue as a powerful decision tool. Throughout the post-World War II period, the fundamental difficulty of allocating a significance (or weighting) to the social and environmental factors which make up that intangible criterion called 'quality of life', remained a serious limitation to systems analysis.

2.6 Social costing of alternative land use impacts

Resources such as landscape, fresh water and the atmosphere are not owned by individuals but are available to the public at large. In the past three decades there has been a growing awareness of the value of these *public amenity resources* and of the necessity to put a monetary or social value on their maintenance. For such amenity resources, the normal law of supply and demand cannot be used to fix their price. Nor is the marketplace, without intervention by government, the appropriate means for determining their value (what users will pay).

The land use options and the alternative uses of what are called *common property resources* (CPRs), must thus be considered in a social framework where their conservation aims to provide the *greatest benefit to the largest proportion of society or the longest period*. This objective has become a goal of sustainable development, or rather, a cornerstone on which sustainable societies can be built. In planning land use, at least three types of costs to society must be considered. One obvious cost is that of regenerating or rehabilitating an area after land degradation such as erosion control, revegetation or detoxification of salinity, heavy metal contamination or pesticides. Another is the initial cost of preventing damage to the landscape; for example, runoff control earthworks, protective tree planting, or more expensive procedures to eliminate the use of toxic but cheaper chemicals. A third cost is the unaccounted *disbenefit* of losing access to land used by the community (e.g. for recreation and tourism) because of environmental disruption or degradation. Since this cost is difficult to estimate it is often excluded when land use planning options are costed. Even modern sophisticated accounting systems do not properly represent what the loss of environmental quality costs the community. The reason for this is simple: the value of these amenities cannot be measured in simple monetary terms.

The growth in the demand for recreation opportunities has been double the growth in population in affluent countries approaching zero population growth (ZPG). As a result, land use planners and land managers are putting more economic value on recreation use, thus competing for landscape and space with traditional production-oriented land uses such as agriculture, forestry and grazing. Serious incompatibilities arise: for instance, when the areas favoured by bushwalkers and birdwatchers border on farming areas where dust, noise, chemicals and disturbance in general are integral aspects of such production systems. The recreation value of the land then becomes the agricultural replacement value, i.e. what wilderness would fetch at auction if it could be converted to cultivation. The same applies to costing wilderness options in landscapes with mining or forestry potential.

Apart from the tourism values of the unique fauna and flora of this continent, the scientific importance of the Australian gene pool warrants special attention. While admirable progress has been made with expanding the network of Australian national parks and World Heritage areas, the conservation status of our flora remains unsatisfactory. Table 2.1 indicates the extent to which known rare and endangered plant species are protected in reserves:

Table 2.1 Number of all reserved flora species, by conservation status and adequacy of reservation, Australia

Adequacy of reservation	*Conservation status* Endangered	*Vulnerable*	*Rare*	*Poorly known*	*Total*	*% of total of reserved species*
Adequately reserved	5	51	263	9	327	19.1
Inadequately reserved	31	145	45	10	231	13.4
Reserved population size unknown	15	194	745	205	1,160	67.5
Total	51	390	1,054	224	1,719	100.0
Per cent of total of reserved species	3.0	22.7	61.3	13.0	100.0	

Note: All species are known to be represented within a National Park or other proclaimed reserve.
Source: Briggs and Leigh, 1988

In the case of fauna, Australia has a record of having already lost 30 species, and the habitat of many remaining species is rapidly being degraded or totally eliminated by development in some form. The status of our vertebrate fauna is shown in Table 2.2.

Table 2.2 Summary of extinct, endangered and vulnerable vertebrate species, 1990

Vertebrate	*Extinct*		*Endangered*		*Vulnerable*	
	Species	*Subspecies*	*Species*	*Subspecies*	*Species*	*Subspecies*
Amphibians	0	0	7	0	2	0
Birds	10	11	11	13	15	12
Mammals	20	1	26	2	17	1
Reptiles	0	0	6	0	14	1
Fish	0	0	7	0	6	0
Total	30	12	57	15	54	14

Source: Australia and New Zealand Environment and Conservation Council, 1991

2.7 Dependence on land — the great awakening

This chapter considers some biophysical and social elements of the land degradation problem worldwide. Brown and Wolf (1984) coined the phrase 'the quiet crisis' to describe land degradation from soil erosion. The phrase is apt because, unlike more spectacular or emotive environmental crises such as deforestation, whaling or oil spills, the depletion of our global topsoils has gone virtually unnoticed by western politicians right up to the 1980s. It is not that their technical advisers had not brought the scale of the problem to their notice, but rather that the political mileage to be gained by investing in corrective action was deemed as insignificant.

The land degradation problem is very widespread, difficult to recognise in its early stages, and complex in its processes. It is generally an insidious phenomenon which vitally influences humankind's chances of global survival on the only planet known to offer conditions for life as we know it. Such a statement may seem dramatic, but although a small part of human nutrition does come from the ocean fish stocks, the great bulk of our food supply is grown in the very thin layer of topsoil that caps the land surface of the globe. Blaikie and Brookfield (1987) have provided a comprehensive overview of the relations between degradation of landscapes and social progress. They explain the fundamental importance of matching the intensity of land use, notably for annual cropping, to the inherent or enhanced '*capability*' of each portion of land. Theoretically this matching ensures the maintenance of productivity and the stability of the landscape as a whole. In practice, the combination of population pressure, ignorance, greed and political corruption means that only a very small percentage of global lands are used according to their capability or capacity to produce safely in the long term.

The 'great awakening' of national leaders and international coordinators such as the United Nations Organisation's Food and Agricultural Organisation (FAO) to the survival values of soil conservation has occurred only in recent decades. The poor success of World Bank-funded food production projects and the failure of Keynesian economics in maintaining soil resources have contributed to a new realisation of the urgent need for the kind of land use planning and management that value sustainability as an essential ingredient in land use decision-making.

One aspect of land use that has played a central role in the global sustainability debate is deforestation. Whether it is rainforest clearing in Amazonia or brigalow clearing in Queensland, the process has been motivated by a desire to give land to more settlers, even if it destabilises microclimate, erosion, hydrology and the landscape. Today's debate focuses on how human interference with landscape processes can be planned to have minimal effect on the land's future productivity. Whether land degrades under imposed management depends on how natural and human forces cancel each other out. Blaikie and Brookfield (1987) suggest the following simple formula as a guide:

Net degradation = (natural degradation + human inteference) –
(natural restoration + restorative management)

This relationship suggests that management can be either detrimental or beneficial, and that nature uses both regenerative and erosive processes in ecosystem functioning. The 'awakening' mentioned above refers to a realisation by land use planners (and recently by some political decision-makers) that it is the *balance* between these positive and negative forces that determines the biological sustainability of a managed landscape.

Despite all the dire warnings about the degradation of Australian agricultural resources in recent years Davidson (1987) stands like a shining beacon of optimism over a landscape of exploitation. Davidson represents those who continue to ask 'What crisis in land degradation?'. He maintains that there is evidence that most Australian cropping soils are more fertile today than they have ever been. The addition of superphosphate and trace elements combined with the introduction of subterranean clover (and more recently medics) have improved both the chemical fertility and the physical structure of most cultivated soils. Davidson also believes that the destruction of the rabbit and the introduction of legume pastures has reduced soil erosion by both wind and water. He suggests that dry-land salinity caused by clearing trees was inevitable if the land concerned was to be of any use for agriculture and that such salinity is limited to a small proportion of Australia's cropland. Soil acidity he accepts as reversible by well known control methods.

Referring to the 1980s, a decade during which most commentators identified a serious decline in agricultural producers' terms of trade, Davidson states that while income declined in the first half of that decade, the cash surpluses and gross margins of many enterprises were positive, and that overall, the average real income earned by farmers during this period is similar to that in the decade 1954–64. He claims that both the Bureau of Agricultural Economics and the National Soil Conservation Program give unsubstantiated pessimistic evaluations of the farmers' economic position and the condition of the agricultural land.

2.8 Managing degradation

If degradation is defined as a decrease in land capability, then land management encompasses the directed processes by which capability is affected. Thus sustainable land management entails those activities which maintain or even improve land capability. Provided the natural environment in which production systems are practised suit those systems (i.e. remain ecologically stable), simple land use systems can be, or can become, permanent. Primitive forms of shifting cultivation and rotational grazing can be maintained, provided that the intensity of use and return of nutrients approximates the natural situation. These elementary agricultural and pastoral systems succeed over long periods because they avoid the need for high inputs (e.g. fertiliser) by harvesting at a rate appropriate to the land's natural ability to restore or rejuvenate its yield capacity.

Shifting cultivation in the form of slash-and-burn jungle gardens was used for thousands of years by many tropical rainforest societies before population pressure shortened the rotation period, i.e. a more rapid return to the same plot after the previous cropping period. This can be seen in Brazil today where the Amazon Indians are forced to return in three years, which is not long enough for the natural soil fertility processes to rejuvenate the highly leached acidic forest soils.

A useful Australian example of managing degradation is the 'fire-stick farming' of the Myall Aborigines. This system worked because there were only low numbers of humans, and burning was not followed by heavy grazing pressure each year immediately after the fire. The system thus managed degradation by limiting the removal of nutrients to a rate within the capability of those infertile semi-arid ecosystems.

Land management must take account of both the *onsite* (e.g. fertility loss) effects of degradation and the *offsite* or downstream effects (e.g. silting, salination, flooding). This matter will be dealt with in detail under Water allocation and quality control (Section 5.3), but its relevance here is that land use planners must consider upstream and downstream causal relations when devising plans for sustainable management. Both upstream and downstream, the basic ecological essentials of sustainability are controlling soil loss, nutrient loss, toxin buildup and soil structure maintenance.

In managing degraded land, the relative importance of different forms of degradation must be distinguished. Eckersley (1989) cites the Australian Institute of Foresters analysis of the costs of various forms of land degradation which placed nutrient depletion at the top of the list. In comparing forms of degradation, the analysis suggests that for every dollar that dryland salinity costs the nation, wind and water erosion costs $5, soil acidification costs $25, soil structure decline costs $125 and plant nutrient losses cost $625. While such estimates cannot be precise, even with a variation of 50% the loss of fertility in the form of NPK (nitrogen, phosphorus, potassium) significantly overshadows the national loss due to other types of degradation. In Table 2.3, as an example of the extent of nutrient removal from the Queensland grain crop, Asher (1993) presents figures based on Bruce's (1990) data:

Table 2.3 Nutrient removal in wheat and barley grain in Queensland in 1988

	Nutrient removal (t)			
Plant nutrient	*Wheat*	*Barley*	*Total*	*Fertiliser additions (t)*
Nitrogen	53,340	10,500	63,840	11,289
Phosphorus	5,511	1,085	6,596	2,062
Potassium	6,597	1,400	7,979	69

Source: Asher, 1993; Bruce, 1990

An important case study in managing degradation is the joint study on 20 groups of catchments on the Queensland north coast. This study arose from the federal government's concern about the impact of agricultural practices on both the Barrier Reef and the coastal zone ecosystems including the important mangrove swamps. The findings of the study can be extrapolated to a range of other coastal zones and consequently are of national significance in ESD planning. In summary, the study found the following (Moss et al., 1993):

- The combined east coast of Queensland delivered close to 15M t of silt into the marine ecosystems each year, with the Burdekin-Haughton catchments producing the major proportion, i.e. 2.83M t/yr.
- An estimated 77,000 t N and 11,000 t P were lost annually from Queensland's coastal catchments. The losses varied by a factor of at least 5 (500% variation) in the northern wet tropics and by at least 100 in the drier catchments further south.
- Most of the sediment and nutrient losses originated from grazingland. The point sources (intense localised outfalls) of N and P represent only a small proportion of total N and P losses, except where large population centres occur, notably in the Brisbane-Gold Coast area.
- The loss of sediment, N and P from rural lands (non-urban sources), is estimated to be about 400% higher than losses in the pre-settlement era, indicating that natural siltation while exporting nutrients was of a significantly lower order.
- Predictably the study showed that catchments with high runoff rates (low plant cover and steep cultivations) gave the highest export rates of sediment, N and P per hectare.
- The study emphasised the potential hazards of silt and nutrients to the Barrier Reef itself, recognised as Australia's top tourist attraction. Recommendations centre on minimising soil erosion through integrated catchment management. Improved land management practice on all properties is expected to give the best short-term results in high runoff catchments.

The influence of agricultural practices on coastal and marine ecosystems is likely to become more significant in the near future. As the socio-economic ramifications of this 'downstream' effect of unsustainable agriculture gain political recognition, so the pressure on rural landholders to minimise off-site effects will increase. These externalities are already being reflected in the objectives and strategies of integrated catchment management (ICM) programs in various states. While the Queensland catchments study concentrated on soil and nutrient losses, it is clear that industrial discharges of toxic wastes, agricultural runoff containing pesticides, and the 'red tides' (blooms of phytoplankton) resulting from high nutrient level in runoff, all require more effective monitoring and control than government deemed necessary in the past.

2.9 Fragile marginal land — the danger zone

Terms such as '*fragile*' (or 'sensitive' or 'susceptible') are often used to indicate the tendency of certain ecosystems or land forms to collapse when overused. The fragility referred to usually stems from the steep terrain, deficiency of nutrients or lack of moisture which characterise the particular land. The International Federation of Agricultural Producers (1991) list '*resilient*, *regenerative* and *stable*' among the more useful criteria for identifying sustainable crop production systems. The level of management required to secure such attributes in a system will increase as the 'fragility factors' increase. In other words, deep, level, fertile valley soil with a high organic matter content and good drainage would be deemed inherently resilient, irrespective of the management superimposed on its surface. Similarly, steep, shallow infertile hillside soils have inherent dangers related to their tendency to erode when the natural vegetation cover is disturbed.

Whether, in our quest for sustainability, we refer to degrees of fragility or resilience (like deciding whether the bottle is half empty or half full) is immaterial — what's important is not the terminology but the relativity of the dangers associated with loss of productive capacity. At this point '*fertility*' should mean both the *chemical* (e.g. nutrients, acidity, toxins) and *physical* (e.g. texture, structure) attributes of soil as a plant growth medium. All these factors combine to make soil productive in yield terms and to give the system sustainability, at least in biophysical terms. Managing all these factors simultaneously is a challenge whenever the most cost-effective control of one factor conflicts with the control of another factor; for example, when the leaching of saline salts by heavy irrigation tends to aggravate rather than alleviate acidity problems. Valley bottoms and alluvial plains represent many of the world's most productive and sustainable croplands, despite having to be drained and modified in the initial land use planning phase. Such modified natural wetlands are a good example of how human interference can permanently upgrade land capability.

The term *marginal* has gained common usage meaning low-yielding, high-risk or unreliable in its productivity. Thus the world's marginal croplands are almost invariably semi-arid, or characterised by shallow soils or steep terrain. The original economic interpretation of 'marginal' was land that yielded only enough to cover costs, but it has been overtaken by an environmental emphasis on its ecological significance. Marginal lands are most at risk from degradation for two reasons: (1) they are physically prone to soil loss by water and wind erosion; (2) their low and variable yield cannot justify investment in preventive measures to ensure soil stability. These constraints limit the sustainability of agriculture on marginal lands worldwide, causing the so-called 'desperate ecocide' — a phrase coined to describe ecological damage by rural communities whose only option is to occupy land inherently unsuited to annual cultivation.

Globally, two of the most well-known spheres of marginality are the Sahel (sub-Sahara) in Africa, and the mountain croplands of Asia. The Sahel is a semi-arid belt of unusually drought-prone tribal lands stretching right across central Africa. Apart from the ravages of tribal warfare, this zone epitomises the worst of human conditions anywhere on the globe — indeed the area including Ethiopia, Somalia and Sudan is regarded by the United Nations not as part of the Third World but as a Fourth World. It is defined primarily by its almost total dependence on foreign aid for a dependable food supply. This situation reflects widescale unsustainable agricultural and pastoral systems in which the demands of the world's fastest growing population stretch the inherently low productivity of the landscape far beyond its capability.

In the Australian land situation, marginality is almost always determined by low and unreliable rainfall. Later chapters evaluate the chances of holding the edge of the Australian cropping zone at its present boundaries, and the necessity of rethinking the sustainable boundaries of cropping. With the current emphasis on sustainability in Australian agriculture, it is important to review past practices and assess their long-term effect on both production levels and the soil resource. It is worth noting that most long-term cropping trials have been made in the semi-arid temperate zone from Western Australia, through South Australia to the Wimmera and Mallee of Victoria. The long-term trials concentrated on the effects of (a) fallowing, (b) fertilising, especially with P, (c) ley cropping with different pasture species and lengths of the ley, and (d) rotating cereal crops with legumes. Only in the past two decades have tillage practices been of interest to researchers. Grace and Oades (1994) have summarised the known long-term (over 25 years) trials in Australia, as shown in Table 2.4.

The data in Table 2.4 indicate that Australia's oldest cropping trial is now over 80 years. Taken as a group, these trials are important indications of how organic matter, P and N, are affected by crop rotations, and how soil acidity has been affected. With superphosphate introduced in the 1880s and legume rotations in the 1940s, most of the long-term experiments focused on the effects of these practices on wheat yield. The old Longerenong Permanent Rotation Experiment, started in 1913, still uses the original varieties such as Ghurka wheat and Swan oats. This benchmark trial gave an average annual yield from continuous wheat of 603 kg/ha for the 70 years starting in 1918. Over that period there was a very slight increase in yield per hectare, but it is important to note that while a fallow period after wheat resulted in a 300% yield increase, this system eventually caused the most serious decline in yield: 12.6 kg/ha/yr. Over the entire period, rotations which included a ley crop (pasture for 1–3 years) gave the best yields, averaging 2,371 kg/ha/yr.

It is not appropriate here to present details of each of the long-term trials. In essence they demonstrate a rundown in organic matter, and the need to maintain the P and N exported in the annual grain crop. Of more recent concern is the correlation between the loss of organic matter (as reflected in organic carbon measurements) and the decline in soil structure, infiltration and water-holding capacity.

Oades made a detailed analysis of chemistry of soil organic matter and found virtually no difference in the chemistry of the organic matter taken from soils under widely varying rotations, but

Table 2.4 Long-term agronomic trials in Australia

Location	*Type*[1]	*Commenced*[2] *(year)*	*Soil type*[3]	*Rain (mm)*
Rutherglen (Vic)	C (RR1)	1913 (1973)	Natrixeralf	593
Rutherglen (Vic)	C (GM)	1913 n.a.	Natrixeralf	593
Rutherglen (Vic)	P	1914	Natrixeralf	593
Werribee (Vic)	C	1914 n.a.	Rhodoxeralf	541
Werribee (Vic)	C (GM)	1914 n.a.	Rhodoxeralf	541
Dooen (Vic)	C (LR1)	1917	Torrert	425
Dooen (Vic)	C (GM)	1917 n.a.	Torrert	425
Kybybolite (SA)	P	1919 (1958)	Natrustalf	549
Glen Innes (NSW)	C	1921	Natrustalf	796
Urrbrae (SA)	P (P1)	1925 (1955)[4]	Rhodoxeralf	627
Urrbrae (SA)	C (C1)	1926	Rhodoxeralf	627
Walpeup (Vic)	C (MR1)	1936 (1973)	Calciorthid	330
Walpeup (Vic)	C (MM1)	1940	Calciorthid	330
Rutherglen (Vic)	P	1946	Natrixeralf	593
Walpeup (Vic)	C (MC4)	1946 (1971)	Calciorthid	330
Walpeup (Vic)	C (MR2)	1946 (1975)	Calciorthid	330
Dooen (Vic)	C (LR2)	1946 (1981)	Pellustert	425
Walpeup (Vic)	C (MR3)	1951 (1980)	Calciorthid	330
Tamworth (NSW)	C	1966	Pellustert	690
Merredin (WA)	C (66M29)	1966	Paleustalf	302
Newdegate (WA)	C (67N4)	1967	Natrustalf	390
Chapman (WA)	C (67C13)	1967	Palexeralf	412
Esperance (WA)	C (68E5)	1968	Paleaquult	496
Salmon Gums (WA)	C (68SG5)	1968	Calciorthid	337
Warwick (Qld)	C	1968	Pellustert	717

[1] P = continuous pasture trial, C = crop rotation trial; location specific identifier in parentheses
[2] termination year in parentheses
[3] Soil taxonomy approximation only (Soil Survey Staff, 1991)
[4] approximate termination date
n.a. = not available
Source: Grace & Oades, 1994

also found that the amount of organic matter present varied by 300% in different systems. Oades concluded that the chemistry (type) of organic matter was determined largely by the soil microbes present, not by agronomic practices, and suggests that the accumulation of fine organic particles (as humus) apparently take several decades to be detectable as an integral characteristic of the profile, actually changing the soil type.

The long-term trials gave little attention to soil biology, but Barley (1959) demonstrated that eliminating the ley crop from the rotation reduced earthworm populations to below 10% of previous levels. The future stability and productivity of Australia's marginal cropping zone will greatly depend on the research and extension of how the soil processes function under alternative cropping techniques. One outcome may be that the frontiers of dryland wheat production will have to be pulled back if technology cannot maintain the soil.

Chapter 3

AUSTRALIAN LAND RESOURCES: CAPACITY OF THE GREAT SOUTH LAND

This chapter examines how the concepts introduced in Chapters 1 and 2 apply to the contemporary Australian situation. The initial optimism of the settlers and the government is contrasted with the instability caused by applying European methods 'Down Under' to landscapes not suited to such traditional rural technology. The chapter analyses the biophysical problems of land degradation within the apparently untenable financial framework in which landholders must operate. The chapter ends with the essential concept of reconciling the economic and ecological requirements of sustainable rural production.

3.1 The vision splendid — an El Dorado Down Under

No purpose is served by yet another description of this vast and ancient land with its unique fauna and flora and its origins in the even more ancient super-continent of Gondwana. Those who wish to study Australia's natural treasures are referred to the ABC Video Bicentennial Series (1988) *Nature of Australia* — an exceptionally incisive account of the ecosystems of this remarkable continent.

Of importance to this chapter are the potential and limitations of Australian landscapes as managed resources for agriculture, pastoralism, mining, forestry and tourism. Almost 70% of the land suffers from severe and unpredictable moisture deficiency. Most soils, except for relatively small areas of recent volcanic earth, are infertile and have poor physical conditions. The rainfall limitation constrains agriculture to about 10% of the nation's land area although just over 65% of the total land is used for rural production including extensive grazing by domestic stock. The productive capacity (in raw materials and amenities) of such a dry continent depends on two elements of land management: (1) the application of systems that use the resources in their natural state efficiently and sustainably; (2) the development of irrigation schemes that make optimal use of the river waters, particularly on the eastern seaboard. The extent and effective use of each of the major resources is considered in Chapter 6 (cropping soils), Chapter 7 (native pasture) and Chapter 8 (forests and minerals); this chapter examines the limitations, potential, hazards and benefits of the Australian resource base in an attempt to clarify the alternative optimistic/pessimistic visions of the nation's resource-based future.

Australia's history since white settlement reflects a continuum of triumphs and failures, winners and losers, wealth and bankruptcy, following in quick succession behind a totally unpredictable procession of floods, droughts, bushfires, insect plagues and weed infestations. The romance of pioneers opening up new country is overshadowed by a higher rate of failures of settlement than occurred in other colonies during the same era. This failure was caused largely by the power of untamed nature to overwhelm governments' and landholders' early schemes for managing their new and poorly understood landscapes.

It is worth noting that soon after landing at Port Jackson (Sydney), Governor Phillip had 9 acres cleared for planting the colony's first wheat crop. It failed, but after several more attempts a grant of 30 acres was made to a convict farmer, '*as an experiment to ascertain how much land was necessary to maintain a farmer and his family*' (Wadham and Wood, 1939). Australia's first land capability studies had begun, having already established that neither grain nor vegetables would grow in the Sydney sandstone soil without manure. By 1806 Governor King had recognised the limitations of both soil and climate in the Parramatta area and his correspondence warrants quoting here as it represents the first land evaluation report on the new continent:

> It is not my intention to discourage the growth of that valuable grain, but I do not think it is safe for the settlement to rely wholly on wheat for general support ... It is soon destroyed in the fields by blights, rust, smut and the caterpillar; also in this climate by fire, and in the stack by weevil; and by corn moths when in the granary, added to which, when the wheat is continually sown on the same land, it impoverishes so much that, if the crop is not destroyed by any of the above common evils, the produce will be small, and by no means equal to the expense of raising it, as the same expense must be used on worn-out lands, that will not produce more than eight bushels of wheat an acre, as upon land that will produce twenty-four. (Wadham and Wood, 1939)

Thus in a single communique King encapsulates the original concepts of land capability, sustainability and marginality — both economic and ecological.

Within the confines of the climate, soil and vegetation of a new country, the general pattern of land use is usually determined primarily by identifying export commodities (e.g. cotton and tobacco in the US) and by the choice of a land tenure system for rural settlement and production. In the words of an anonymous historian, 'we look in vain for anything logical or orderly in the official control of settlement', so we are left with the choice of saleable products (wool, timber, gold and wheat) as the prime determinant of land use patterns. Thus was born a nation forever tied to cycles of boom and bust, of achievement and heartbreak, of giving it a go even against long odds. The natural vagaries of the environment bred optimists and pessimists, idealists and realists, each one holding a worldview that reflected their experiences in what happened to be their personal segment of the Great Game of Chance played out in the Great South Land.

Lines (1991) has given a masterly overview of the exploitation of Australia's natural resources using the apt title *Taming the Great South Land*. One of Lines' themes is the way in which Australians viewed their land purely as a wealth-generating resource. Except for a few important visionaries who warned of impending ecological collapse, governments, politicians, administrators, rural producer leaders and individual landholders based their sense of values on what appeared to be endless and inexhaustible land resources. This view was no doubt heightened by the vast distances unmatched by the proportions of the countries of origin of the settlers faced with exploiting ecosystems of which the world had no knowledge or understanding.

3.2 Settling new lands, unsettling old ecosystems

By the late 1820s 2 million ha of the young colony of New South Wales had been surveyed for a farming population of 70,000. Although settlement beyond the surveyed boundaries was forbidden, the squatters had moved their flocks and herds out to 550 km from Sydney by 1835. Thomas Mitchell in 1836 reported on Western Victoria in glowing terms '... Every day we passed over land, which, for natural fertility and beauty, could scarcely be surpassed; over streams of unfailing abundance, and plains covered with the richest pasturage' (Lines, 1991). This was Mitchell's Australia Felix, a land of seemingly endless rural resources, soon to be doubly lucky when gold was discovered in the 1850s.

Between agriculture, pastoralism and mining, the landscape was irrevocably changed, first by removing trees, then by ploughing sloping land, and finally by overgrazing native pastures. The enlightenment which colonisation was meant to bring through its conquest of nature produced a shallow temporary society built on wasting foundations and, in an ecological sense, living on borrowed time. The great abundance of nature's harvest was gained at the expense of ecological stability.

The early effects of agriculture are vividly described by Bolton (1981) under his evocative title *Spoils and Spoilers* reflecting the effects of exploitation. By the turn of the century, the erosion by wind and water in the Western Division of New South Wales was so serious that a Royal Enquiry was held to determine the causes and recommend corrective action. This enquiry was a milestone in Australian land use and represents the first major case study of the extensive use of land beyond its capability. To this day, sheep numbers in the Western Division have never again even approached the stocking rates of the late 1800s. Figure 3.1 shows how drought has been a natural determinant of productivity for nearly 40% of all seasons since the 1860s.

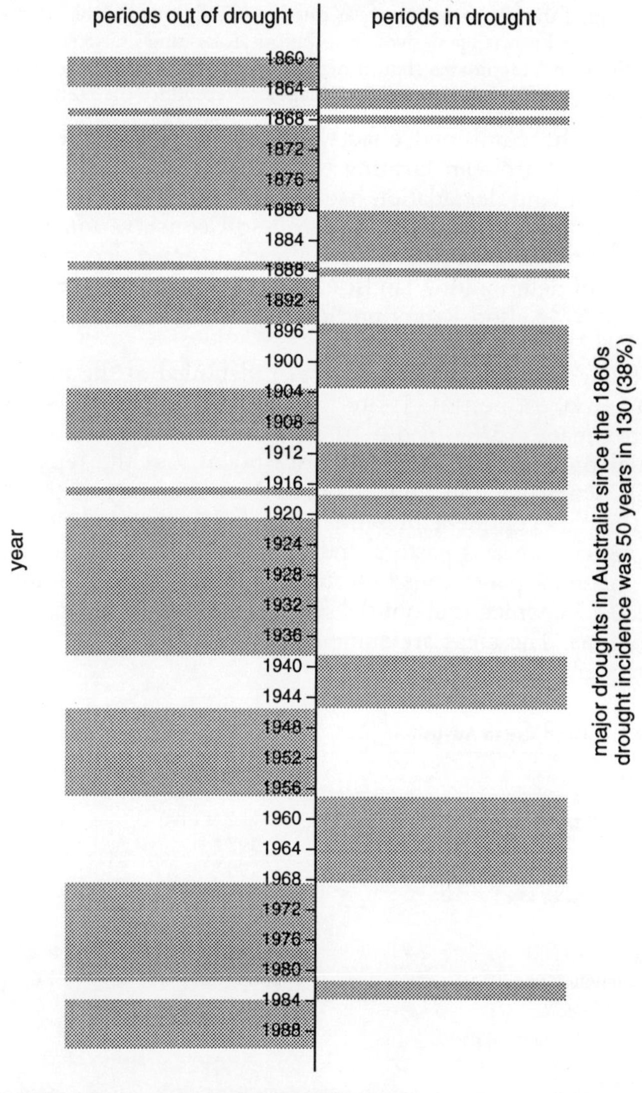

Figure 3.1 Distribution of Australian drought periods
Dept. of the Arts, Sport, the Environment and Territories, 1991

While the rabbit plague was blamed for much of the degradation of Australia's farmland in the early 1900s, the major disruptions of the ecosystems arose from rural production and mining operations. In Messer's (1987) view:

> All accounts of historical land degradation in Australia draw attention to the lack of resilience of much of the land and its consequent sensitivity under intervention. An essential correlate is the widespread lack of understanding of the ecological characteristics of the land and the slow and subtle but inevitable nature of the different forms of degradation. It is generally accepted that this ignorance was due historically to a combination of inappropriate knowledge derived from European farming practices, and cultural values and ideologies that either devalued nature or overestimated the capacity of natural ecosystems to adapt and reproduce themselves.

By 1900 the combined effects of uncontrolled timber clearing, unadapted European farming techniques, undersized settlement blocks and land degradation had reduced production from cropland and from natural grazing. By 1930 soil conservation legislation and property amalgamation were already deemed necessary to stem the tide of deteriorating landscapes. Yet it took until 1980 to set up the first Vegetation Retention Scheme (S.A.) requiring consent for clearing.

Today land degradation has been identified as the most significant environmental issue in Australia (Roberts, 1987). Approximately 51% of our rural lands are degraded; the loss of annual production is estimated at $600M and the repair cost at $1.6B (Dumsday and Edwards, 1984). A summary of the proportions of this continent presently used for various industries shows that agriculture and pastoral industries use about 61% of our land area, national parks constitute 5.3%, forestry about 3% and urban areas 0.13% which is about the same as the area under active mining operations. The areas are summed up in Table 3.1 for agriculture, forestry and parks:

Table 3.1 Land use in Australia

Land use	*As at*	*Area (Mha)*
Agricultural establishments	31 March 1990	466.6
Crops	1989–90	17.0
Sown pastures and grasses	1989–90	30.9
Grazing/fallow or idle	1989–90	418.7
Forestry		
Native forest	30 June 1990	40.8
Plantation	31 March 1990	1.0
Protected areas		
Wilderness, national parks, conservation reserves, etc	1989	36.5
Total Australian land area		768.0

Source: Castles, 1992

In ecological terms the original Australian ecosystems and the modern agricultural systems are best compared on the basis of their diversity or simplicity. Seven important ecological differences can be identified:

1 Natural systems capture, convert and store energy from the sun, while most agricultural systems consume non-renewable fossil fuel.
2 Natural systems consume carbon dioxide and produce oxygen while technical systems do the opposite.
3 Natural systems create and conserve soil while cropping systems deplete and diminish fertile soils.
4 Natural systems purify water and usually release it gradually while most intensive agriculture contaminates water and causes excessive runoff.
5 Natural systems provide a range of comprehensive wildlife habitats while agriculture tends to reduce habitat range and extent.
6 Natural systems cycle and detoxify their products of decomposition while agricultural systems produce pollutants that are costly to detoxify or remove.
7 Natural systems have the capacity of self-renewal and maintenance while agricultural systems require continued expensive inputs to remain productive.

3.3 Current resource problems — physical and economic

Sustainability in Australia's land resource-based industries differs between the industries. Mining, which is based on non-renewable minerals and energy, has environmental problems related to rehabilitation, acid-mine drainage and, to a lesser extent, air pollution. It is an extractive industry in which the mining techniques and the rate of extraction depend largely on international market forces. Forestry, as a resource industry, depends on renewable timber resources, harvested at a rate appropriate to their regenerative capacity. Forestry extraction techniques can be environmentally friendly within the limits of felling procedures and haulage machinery. Roads, drainage, erosion control, wildlife protection and control burning all contribute to a system which approaches sustainable forestry if applied correctly.

As referred to earlier, one of the major causes of the Australian landscape's induced instability has been the clearing and changing of the vegetation cover. This change has been important from several points of view: biodiversity (as habitat), stability (as erosion control) and amenity (as recreation area). An attempt to estimate the proportion of Australia's land area affected by the six major grades (or intensities) of vegetation modification over the past 200 years is given in Table 3.2.

Table 3.2 Modification to Australian natural vegetation 1788–1980

	Area of Australia %	
• Unmodified natural vegetation — no apparent use by Europeans	30	Increasing degree of modification ↓
• Natural vegetation altered by extensive grazing or forestry and/or recreation activities	47	
• Natural vegetation modified by grazing and partly replaced by introduced species	7	
• Clearing or partial clearing of upper stratum and widespread modification and replacement of lower stratum with introduced species	6	
• Cultural vegetation only	10	
• No continuous vegetation (includes urban and commercial areas, opencut mining and man-made lakes)	<0.5	
Total	100	

Source: Department of Home Affairs and Environment, 1981

Tourism is a resource industry of a different type. While not being extractive, it uses landscapes as its base and is very sensitive to their condition or 'naturalness'. The land resource problems in tourism are mainly those of competing uses, especially uses that deface or degrade the natural beauty of the landscape. Sustainable land use for tourism thus depends on controlling problems arising from the numbers of tourists and the means by which they are transported, fed and accommodated. In addition, recreational activities such as walking in rainforests, on coral reefs or up alpine mountain trails need strict control if the environmental quality enjoyed by tourists is to be maintained. As Australia's fastest growing industry in terms of foreign income generated, tourism as an alternative in multiple-use land planning requires special consideration in Australia's *sustainable future*.

Agriculture and pastoralism are currently associated with Australia's most serious environmental problems, notably the various forms of land degradation. The *physical problems* may be divided into two main groups: the first encompasses the obvious physical and chemical problems such as wind and water erosion, dryland salinity and irrigation salinity. For example, 35% of the vast inland rangelands (grazinglands) are suffering at least moderate degradation from erosion and salinity. This first group of problems is characterised by obvious outward symptoms of change and retrogression, easily observed from a distance. One of the problems of degradation is the 'downstream' effects referred to later in Section 5.3. From a national and social costing point of view, the division of the costs of degradation in Table 3.3 indicates the range of effects both on the farm and downstream:

Table 3.3 Degradation costs and effects

A Onsite Productivity costs	
Output	Costs of product loss due to changes in soil quality or quantity.
Input	Effects of onsite land degradation on cost of production, e.g. increased costs of fencing or dams, replacement of machinery, or indirect cost of compensating for ecological changes, e.g. loss of tree cover for natural predators.
B Offsite externalities costs	
Productivity costs	Product lost or costs increased by external economic activities, e.g. downstream salinity effects, air pollution, regional effects of loss of flora and fauna.
Other costs	Increased costs or lost benefits due to: • damage to water conveyance, irrigation or water treatment facilities • flood damage (e.g. silting of roads) • decreased capacity of water storages • soiling and materials damage • increased navigation hazards • reduced availability or quality of natural recreation facilities • impaired aesthetic characteristics • lost ecological diversity

Source: Chisholm and Dumsday, 1987

The second group of degradation problems concerns the more subtle and hard-to-detect problems occurring in the soil and in the biological populations of fauna and flora. Two specific processes — soil structural decline and the loss of organic matter — seriously reduce the productive capacity of cropping soils and jeopardise the stability or *erosion-resistance* of many soil types. These changes reduce the infiltration of rainwater, simultaneously increasing runoff and soil loss. They reduce the water-holding capacity of the soil and thereby reduce crop growth and yield. At the same time, the reduction of the soil's organic matter reduces its fertility and biological activity. This second group also includes soil acidification, caused mainly by high nitrogen in the sheep/wheat systems of the temperate zone. Acidification can be easily remedied by adding agricultural lime, but the system's effect of lowering soil pH to below 5.0 over vast areas of the winter rainfall region is having severe economic consequences in the southern states. All rural land use planning must take into account the very limited extent of highly productive soils in Australia, as shown in Figure 3.2.

The *economic problems* of the rural industries have changed as farming has progressed from the pioneering era. The National Farmers' Federation suggests that farmers have evolved from their initial stance of *settlement and survival* to *settlement and exploitation,* and on to today's *settlement and sustainability*. While the boundaries between these stages are vague, the basic changes in goals and attitudes have changed according to circumstances. First the fight against

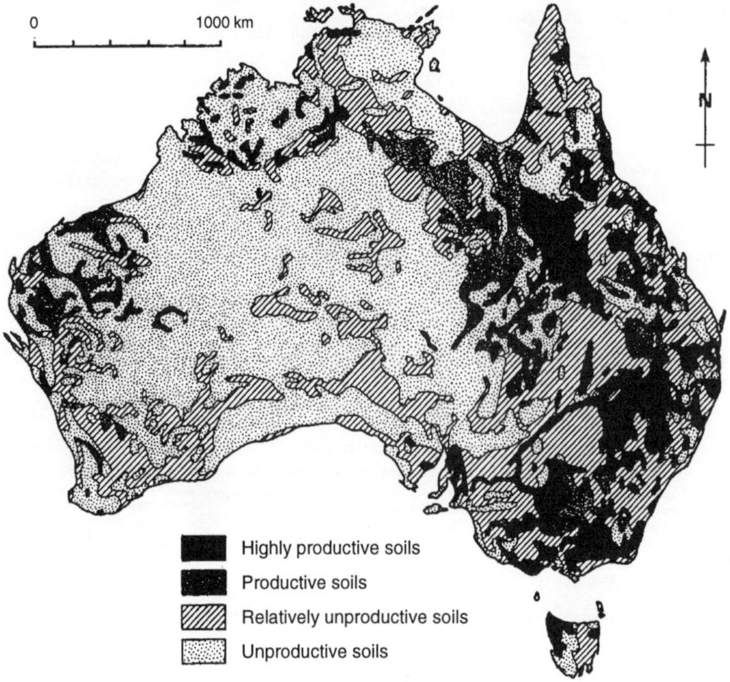

Figure 3.2 Geographical distribution of highly productive Australian soils
Cocks, 1992

nature, then the technology available to cash in on broadacre harvesting, and now the growing realisation that without a change in farming practices, rural industries probably do not have a long-term future.

Agricultural economics and agricultural practices are mutually dependent and affect each other in a continuous series of actions and reactions. The complexities of the economics of sustainable agriculture are dealt with in Chapter 17; at this point only the interaction between biological and economic sustainability need be considered. Today Australian agriculture is suffering from the long-term side-effects of several of the rural industries' 'essential' inputs into progressive agriculture — superphosphate, subterranean clover, tree clearing, fine-tilth cultivation, irrigation, total weed control and chemical pest control. The effects of these operations depend largely on the soil type and the intensity of use of each production factor. Superphosphate raises yield but may lower soil pH; subterranean clover in rotation with wheat may aggravate soil acidity; clearing in catchments with sedimentary soils (from sandstone or marine deposits) may cause serious salinity problems; fine cultivation was formerly recommended but is now associated with structure breakdown and surface sealing; irrigation was to be the key to the success of the Murray-Darling scheme but is now recognised as the cause of widespread salinity;

weed control with herbicides replaced regular cultivation but is causing problems of drift to nontarget plants and of runoff affecting riverine vegetation; pesticides have controlled insects in the short term but are now suspected of causing health problems.

In addition to these negative effects, agricultural activities have reduced wildlife populations with random poisoning and by significantly reducing habitat. Cattle and sheep grazing in the vast inland is today associated with major habitat change and increasing threats to rare fauna species.The challenge to modern rural industries is to find ways to protect the environment while maintaining export earnings. In other words, sustainable agriculture must develop systems that increase production but maintain the quality of soil, water and air. To meet these objectives, the National Association for Sustainable Agriculture of Australia (NASAA) was formed during the 1980s. Although this organisation is heavily biased toward *organic* (non-chemical) agriculture, many research organisations are now focusing on *permanence rather than yield* as the prime criterion of appropriate modern agriculture.

Research into better crop varieties, improved fertilising, more effective tillage, weed control and pest management have led to improvements in the yield of grain crops in Australia. However, if our improvement in yield is compared with our international competitors it is found that over the 35-year period to 1986 Australia's rate of increase was one of the lowest, as Hamblin (1991) shows in Table 3.4.

Table 3.4 Mean yields and rates of increase in cereal yields in various climates over 1950–1986

Climatic zone	*36-year mean yield (tonne/ha)*	*Current yield (tonne/ha)*	*Rate of increase (kg/ha/year)*
Temperate, maritime			
UK	4.0	6.0	94
West Germany	3.7	5.3	84
France	3.6	5.2	114
Mid-latitude, humid			
Eastern USA	3.3	4.7	84
Argentina	1.8	2.5	42
Canada	1.9	2.4	34
Semi-arid, Mediterranean			
Australia	1.2	1.6	12
Syria	0.8	1.2	14
Morocco	0.8	1.5	13
Libya	0.4	0.7	14
Turkey	1.5	2.0	33
Wet tropics			
Indonesia	1.8	3.5	68
Philippines	1.3	1.9	29
Thailand	1.8	2.0	24

Source: Hamblin, 1991)

The difficulty of relating crop yield trends to soil degradation now warrants closer examination. Hamblin and Kyneur (1993) recently examined wheat yields and soil fertility for the period 1950–1991.

Compared to other wheat-producing countries, Australia had one of the lowest rates of increase in wheat yield. The study indicated that the average yield increase in Australia was no more than 8 kg/ha/yr. In districts where ley crops and legume pastures had been implemented, the increase in yield was nearly 40 kg/ha/yr. The study points out that in some districts even where improved cultivars, better tillage practices and improved pest control had been applied, the wheat yield remained constant or declined slightly. This result is apparently due to fertility decline, a conclusion supported by the fact that in Australia's 210 wheat-producing shires, about a third (36%) of shires reflect yield increases below 9 kg. It can be assumed that the significant potential promised by new wheats is being constrained by soil limitations.

From an economic viewpoint it is important to note that Hamblin and Kyneur found that in over 25% of all wheat shires an increase of over 20 kg/ha had been attained as a result of larger machinery which gave more timely sowing, greater weed control and better fertiliser application. Almost invariably those shires returning declining yields had no or little integration of legume rotations. In many cases the use of monocultures and frequent cultivation was associated with declining soil organic matter. The lack of legumes was associated with reduced soil nitrogen and resultant lower wheat protein content. Earlier research had shown that stubble retention increased water use efficiency by 5–12%. Weed control had given 5–15% better water use, minimum tillage gave 5–10% and fertiliser application could give up to 50% better water use. The financial implications of these findings are serious in the present national and international grain export situation.

The long-term trends in wheat yields referred to above should be expanded in terms of the relationship between sustainable yields and soil fertility. As a resource management problem peculiar to the Australian soil situation, this relationship is fundamental. The early decline in wheat yields from 1870 to 1900 in the southern states is directly attributed to the rundown in natural fertility of newly settled lands under continuous cultivation (Williams, 1989). Fertiliser application, new varieties and fallowing returned yields to their 1870s level by the 1940s. Since that time, however, average yields have not risen above 1.5 t/ha, which contrasts markedly with the US achievement of doubling yield (from 1.0 t to over 2.2 t over a similar period; Russell, 1973).

It has long been known that on many soil types the available nitrogen (N) levels are closely related to the age of cultivations (decline is most marked in the first decade). This deficiency can be and has been rectified by incorporating nitrogen-fixing crops in the rotation. However, the fertility problem is more acute in the case of phosphorus (P) because there are no obvious inputs to the soil except through natural weathering. This process may add about 0.15 kg/ha/yr while about 3.0 kg is removed by a crop of 1.0 t. If the maximum input of P for the past 90 years on wheatbelt soils has been about 8.3 kg/ha/yr, the P in the top 1.2 m of soil will presently be between 2000 kg/ha (red brown earths) and 5000 kg/ha (black clays) (Williams, 1989). The phosphorus balance sheet might be as shown in Table 3.5.

Table 3.5 Estimates of cumulative inputs, outputs and storage of P in production of wheat since 1900 (kg/ha)

	Red brown earth	*Black earth*
Inputs:		
Weathering	12	16
Maximum fertiliser P	670	670
Outputs:		
Crop P	200	200
Erosion P*	120	400
Storage:		
0–15 cm	600	1800
15–120 cm	1300	2900
Change:		
in P	+362	+86

* Assumes 5 tonnes/ha/yr erosion. Based on data from McGarity and Storrier (1986).
Source: Williams, 1989

3.4 Integrating land resource ecology with economics

When the Australian edition of *Our Common Future* (Brundtland, 1987) was published, it contained a chapter headed 'A Sustainable Future for Australia' by the Commission for the Future. This brief overview represents the nation's first attempt to define sustainable development. Economic *growth* is measured in real income per capita, while *development* simply means *desirable change*. Thus whether growth signifies development depends on the nation's social goals. From this it follows that what constitutes appropriate development depends on what we want for our children, so integrating economics and ecology should aim at improving material well-being while maintaining quality of life through environmental protection.

Meeting both economic and ecological goals requires moving the sustainability debate away from the simplistic '*growth/no growth*' argument to the *appropriate growth* alternative. This means choosing environmentally friendly products, processes and technologies. Is it possible to set guidelines for attaining economic and ecological goals simultaneously? The Commission for the Future describes *economic sustainability* in terms of four attributes:

- *Efficiency* — programs that yield greatest output per unit of input.
- *Investment maintenance* — the application of human, technological and natural resources as an investment base should not run down.

- *Diversification* — a diverse range of inputs and products make programs less vulnerable to shocks caused by unplanned change.
- *External balance* — the value of commodities exported should balance those purchased from outside the economy.

Ecological sustainability, on the other hand, is described by five attributes:

- *Biological diversity* — stability, adaptation and evaluation are conserved by maximising the range of fauna and flora protected.
- *Ecosystem conservation* — the interaction between biotic and abiotic components requires protection of a holistic nature, preserving the internal interdependencies of the system.
- *Interconnectedness* — at national and international level the cause-and-effect relations within the finite globe require consideration.
- *Aversion to risk* — conservative planning avoids the risk of catastrophes from poorly understood factors such as chemicals and climatic change.
- *Scale of impact* — human activities should minimise the scale of influence of technologies on ecosystems.

The challenge to land use planners is to find systems (for production, services and amenities) that give what is termed *best fit* between the above requirements for integrating the goals of financial and biological sustainability. Some economists argue that intergeneration equity depends on the inheritance of total capital stocks — not only the natural stocks, but also the man-made stocks, e.g. infrastructure, communications, amenities. This will be further examined in Chapter 16.

The founding of the International Society for Ecological Economics (ISEE) in 1990 heralded the formal acceptance of the interdependence of these two disciplines and their potential contribution to the solution of environmental conflicts. Ecological economics differs from natural resource economics and environmental economics and represents a new world view. It concerns the links between economic and ecological systems; its economic models are based on sustainability rather than growth; and it examines questions of equity between generations and species. In the words of Proofs (1989):

> Economists are increasingly coming to recognise that the study of human activities on a finite planet, in the long run, requires a different set of concepts to those useful for the economic analysis of households, firms and nation states in the short and medium run. In a complementary way ecologists are ... recognising that economic activity is here to stay, human activities are coming to dominate global ecosystems and ecosystem analysis which does not explicitly include economic activities makes less and less sense.

As a practical vehicle for solving modern resource allocation and management problems, ecological economics, as a discipline in holistic thought, is characterised by at least six major themes:

1 Multiple values and broader notions of value, including aesthetic, social, intrinsic and energy.

2 Integrity and sustainability of ecosystems as essentials to future economic well-being.
3 Decisions that do not discount the future at the expense of future generations.
4 The uncertainty principle, which recognises the fundamental risks of large-scale or irreversible changes in the environment.
5 Pluralism of methods, which includes a range of analytic methods to overcome the shortcomings of a single framework.
6 Underpinning ecological economics with a land ethic.

Much of the philosophy of ecological economics is comprehensively explained by Daly and Cobb (1989).

Australia still has a long way to go in developing procedures which properly account for both economic and ecological costs and benefits. While not of direct concern to agriculture, the Wesley Vale paper pulp dispute in Tasmania in the late 1980s is an important case study in the application of policy for ecologically sustainable development. Wesley Vale highlighted the problems associated with including ecological resource use considerations in the traditional economic development mode of policy formulation in Australia, and is significant for two reasons. First, it confirmed the impact of community opposition to major company developments on state and federal political judgements. Second, it highlighted the fundamental deficiencies of Australian policy-making institutions in dealing with the community's emerging environmental awareness. The processes used to plan, approve and present the pulp mill case to the public was ad hoc and inconsistent. When this was followed by a non-sequential review and finally rejected by federal and state governments, it was clear that both the decision-making mechanism and the scientific information on the case were totally inadequate (Economou, 1992). This case study exposed four major deficiencies in resource allocation decision systems: it emphasised again the inadequacies of the Environmental Impact Statement (EIS) as a means of conflict resolution; it showed up the inherent conflicts within the development/finance section itself; it highlighted the inadequacies of present processes of public participation; and it showed up the shortcomings of belated policy formulation as a reaction to public criticism. This dispute is a benchmark because it involved not only paper mills and environmentalists but emotively charged substances such as organochlorines and dioxin, and drew the agricultural and fishing industries into the conflict. It is also one of the few occasions on which the United Scientists for Environmental Responsibility and Protection (USERP) took a public stand.

One modern concept in planning of resource use and landscape evaluation is what is termed the 'Precautionary Principle', explained in the 1992 agreement between the federal and state governments as follows: 'Where there are threats of serious irreversible environmental damage, lack of full scientific certainty should not be used as a reason for postponing measures to prevent environmental degradation'. In applying the precautionary principle, public and private decisions should be guided by: (a) careful evaluation to avoid, wherever practicable, serious or irreversible damage to the environment, and (b) an assessment of the risk-weighted consequences of various options. This concept of precaution is also encapsulated as Principle #15 of the 1992 Rio Declaration. Christie (1993) recognises that the precautionary principle has several significant influences on sustainable development planning. First, it sets a framework for decision-making in cases of scientific uncertainty. This eliminates the danger of identifying impacts only after the damage, possibly irreversible, has been done. Second, it overturns the need for scientific proof since it precludes the need for conclusive scientific proof. In this case it is sufficient to have reason to assume harmful effects, even where there is no scientific evidence to provide a causal link. The third advantage of the precautionary principle is that it reverses the burden of proof — the onus is on the developer to show that no serious environmental impact will occur. Christie (1993) suggests that it is not surprising that a literal interpretation of the crux of the principle ('reason to assume harmful effects') may lead to claims of imbalance favouring the environment. When this principle is applied, the essential information can be qualitative, not necessarily quantitative, as was the case before its application. The current position encourages nations to respond positively to the Rio Declaration of which the precautionary principle is an important element.

Chapter 4

NATIONAL GOALS AND VALUES: THE FOUNDATION OF PLANNING

Following on from the need for environmental realism in land use planning in Chapter 3, we move now to the concepts of ecologically based agriculture and stewardship, and how these fundamental approaches are allied to some of the ancient values of indigenous peoples in various countries. The chapter deals with how permanent wealth generation depends on resource management that takes account of the inherent capability and stability of each type of landscape.

4.1 Standard of living and ecological agriculture

Decision-makers faced with choices of alternative uses for land resources must consider the needs and values of the society they represent. As population density increases and environmental quality deteriorates, the value attached to clean, healthy and beautiful surroundings tends to rise. This revaluing of the quality of the human habitat leads in turn to a reconsideration of what is meant by the term 'standard of living'. As mentioned in Chapter 1, Australia is expanding its definition of its living standards solely in material terms to embrace several quality of life attributes which relate to environmental factors impinging on our daily lives.

Modern planners are thus obliged to pay heed to enlightened public opinion when allocating land for competing purposes such as agriculture, forestry, mining, tourism and recreation. The balance between these enterprises will be guided by the political pressures favouring production objectives or the provision of amenities for the electorate as a whole. These social choices can be translated into personal values and even into the way the community defines 'success' in individual and group terms. Where farmland or mining leases are sacrificed for use as wilderness or recreation areas, economic output is traded for improved lifestyle in environmental terms.

Besides having the choice of agriculture or no agriculture in a particular location, society can also choose between several ecologically appropriate agricultural options. For example, it can select crop production methods that minimise erosion and the use of chemicals, and thereby reduce environmental impacts to a level that makes agriculture sustainable, healthy and economically viable in the long term.

Taking an ecological view of agriculture entails a holistic approach to the land resources and their protection. It means accepting the landscape as a unified whole — as an ecosystem that encompasses producers, consumers and decomposers, plus flows and cycles of energy, nutrients and minerals. It means appreciating the interdependence of the components of the natural system and how the manager's intervention affects productivity and stability in both physical (e.g. erosion) and biological (e.g. toxic chemicals) terms.

Crucial to the ecological view of agriculture is a recognition of the limits to which systems may be harvested while keeping their potential productive capacity intact in the long term. This is sometimes referred to as 'living gently on the land' and it implies an approach to the land based on a personal philosophy of the *balance of nature*.

4.2 The farmer's view of stewardship

Irrespective of the scientists' and academics' views of what constitutes permanently productive systems, it is the land manager who must establish practical, sound practices. So the farmer's view of what constitutes sustainability for the family farm is the most important perception of all. Farmers seek first a reasonable financial return on their capital and labour. They seek efficient production techniques that do not demand too much physical effort or personal time, and they seek a work environment that is healthy and satisfying to themselves and their family. Financially they aim to operate farming methods that can adapt to risk and changes in markets and weather. Personally and socially they prefer a type of farming that keeps their real estate in good condition, meets their peer groups' expectations in land stewardship, and gives their offspring a sound resource base for future family prosperity.

In working toward these goals for the family farmer, landholders cherish their freedom to act independently, to operate as autonomous entrepreneurs and to work with minimum regulation and interference. They value the right to seek information and assistance when and from whom they choose and to decide for themselves how best to manage the land. How far sustainability can be achieved in a democracy depends greatly on landholders' voluntary acceptance of sound land use practices. One of the forces affecting the way they manage their resources is their attitude toward the ownership of land. Some landholders regard freehold title as a passport to unqualified exploitation of the land they hold — to do with as they please with no responsibility to those who follow on. This attitude has a critical bearing on the significance of 'sustainability' as one of the aims of good farmers, for if we avoid all responsibility toward others in later years, the idea of intergenerational equity has no meaning — and this is reflected in our behaviour and our treatment of the land. Writers such as Wes Jackson (1984) and Wendell Berry (1977) have eloquently brought home this element of permanence to modern agricultural societies. From such personal philosophies has grown the 'sustainability-as-stewardship' school of thought reflected in its many forms by the author elsewhere (Roberts, 1993). The idea of stewardship is based on landholders regarding themselves as tenants or temporary custodians of the nation's resources, rather than as outright owners or as end-users of the land. This long-term unselfish view brings with it a respect for the landscape and the humility associated with frugal living and an appreciation of nature. It is the essence of a personal worldview based on sustainability.

Australian farmers are quick to claim that 'all farmers are conservationists at heart', and that because they depend on their soil for their livelihood they will (of course) look after it as good stewards. Begbie (1989) suggests that for this relationship between income and conservation to hold good, we must also believe that (a) the better farmers treat their soil, the higher their income; (b) that farmers are willing to forgo short-term income for long-term stability; and (c) that farmers know for certain how present agronomic practices will affect their soils in the distant future. Begbie takes to task leaders of organised agriculture for peddling the 'conservation cocky' myth, since evidence to the contrary is spread all over the Australian landscape. He declares that 'A neater definition of unsustainable agriculture would be hard to imagine' than that used by a recent producer group leader: 'a production system which is capable of continuing to meet the demands of consumers as well as farmers into the far future'.

In their personal quest for sustainability, landholders have faced a range of financial constraints beyond their control. In the past decade the following changes have occurred (Emery, 1991):

- ◆ Deregulation, which made funds more readily available and increased competition in the finance industry.
- ◆ Relatively high inflation, which increased rural production costs without taking much if any account of farm performance.
- ◆ World markets and trade practices that made the prices of some major agricultural commodities fluctuate greatly.
- ◆ Unprecedented extremes in interest rates which many rural enterprises were unable to meet; these rates caused substantial loss of equity for some, and financial failure for others.
- ◆ Rapid appreciation of farm land values caused by high prices of commodities and assisted by the availability of funds.

4.3 What can we learn from indigenous people?

Every two years Australian environmental planners meet with their New Zealand and Canadian counterparts to discuss assessment and monitoring issues. Gow (1989) stresses the importance of using as fully as possible the knowledge indigenous people have of natural ecosystems when it comes to assessing development proposals: 'it was noted that while some scientists might think they know everything, the store of knowledge of indigenous people can complement scientific assessment'. Gow reminds us that this indigenous knowledge helps to integrate the fragmentary knowledge that reductionist science tends to generate. While Aborigines may have limited knowledge of commercial agriculture, their understanding of fauna and flora and their reverence for their Earth Mother can contribute considerably to our understanding of our unique biota and our attitudes to the land we depend on for our survival. The concurrence of the values of indigenous cultures with those of modern deep ecology is an important step toward global acceptance of ecologically sustainable lifestyles. Capra (1980) gets to the heart of these fundamental values when he describes how humans should view their relationship with the earth:

> Deep ecology is supported by modern science and in particular by a new systems approach, but it is rooted in a perception of reality that goes beyond the scientific framework to an intuitive awareness of the oneness of all life ... it becomes clear that ecological awareness is truly spiritual.

Such awareness is shared by indigenous peoples in a range of habitats. The well-known credo of Chief Seattle (1854), recently attributed to a more recent author, encapsulates the bonding between humans and nature: 'The land does not belong to man, man belongs to the land ... all things are connected ... the earth is man's mother.' When these values (and those of ancient Taoism) are compared to modern holistic thinking, important common values emerge. The principles of deep ecology (old and new) are summarised as follows:

1 The well-being and flourishing of human and non-human life on earth have value in themselves. These values are independent of the usefulness of the non-human world for human purposes.
2 The richness and diversity of life forms contribute to the realisation of these values and are also values in themselves.
3 Humans have no right to reduce this richness and diversity except to satisfy *vital* needs.
4 The flourishing of human life and cultures is compatible with a substantial decrease of the human population. The flourishing of non-human life requires such a decrease.
5 Present human interference with the non-human world is excessive, and the situation is rapidly worsening.
6 Policies affecting basic economic, technological and ideological structures must be changed so that the resulting state of affairs becomes deeply different from the present.
7 The ideological change is mainly that of appreciating *life quality* rather than adhering to a rising standard of living. There will be a profound awareness of the difference between big and great.
8 Those who subscribe to the foregoing points have an obligation directly or indirectly to try to make the necessary changes.

4.4 Economics and sustainability in Australia

The Working Party on Ecologically Sustainable Development received several contrasting submissions from different sectors, notably the Australian Conservation Foundation and the Business Council of Australia (BCA). While the original context of the idea of sustainability was environmental, the economic context is important. The Business Council's (1990) submission gives a rare insight into how commercial and development interests regard ESD. The following summary presents some of the more significant points made by the BCA.

The concept of ESD is an appealing one for it emphasises the links between development and the environment, and properly managed ESD could offer higher standards of living and of environmental quality. Much economic activity in Australia is completely compatible with the environment, thus making it unnecessary to propose radical changes right across the industrial sector. The question must be asked why our economic system does not deliver ESD right across the industrial spectrum. The answer may be found in four principles of economics which impinge on environmental quality:

1 The removal of unnecessary regulations which reduce the business community's ability to perform well economically and environmentally.
2 The expansion of property rights to encourage accountable use of natural resources through pride of ownership.
3 The realistic pricing of resource assets leading to the full and correct valuing of these assets.
4 The development of market-based approaches to environmental issues to encourage appropriate use of financial incentives.

The BCA believes that all the major environmental problems facing Australia at present can be tackled using a combination of the above economic principles. In this way pollution control, for instance, could be achieved — not by setting arbitrary pollution limits and charges but by putting a real value on the cost of environmental damage caused by pollution, and covering this cost by economic instruments. The incorrect pricing of water plus the subsidisation of marginal farming has, in the view of the BCA, been the prime cause of land degradation. The removal of 'perverse' incentives combined with appropriate tax structures could encourage better land use. In the manufacturing sector, market mechanisms that reward innovative companies for pollution control would greatly encourage factory owners to meet current environmental standards.

The BCA has identified inappropriate incentives which 'distort market signals' in resource-based industries. These conditions encourage a 'misallocation' of resources where market mechanisms are distorted by government intervention. Large-scale land clearing and irrigation salinity are quoted as examples of environmental damage caused by ill-directed clearing subsidies and the underpricing of water to users on publicly funded irrigation schemes. It is suggested that the correct pricing of timber from Crown forests, tax deductibility for mine site rehabilitation and unsubsidised rail transport would all contribute to environmentally beneficial resource management.

The BCA suggests that, except for fauna and flora that do not require protective legislation, other resources could be more efficiently used and protected if market-based solutions were sought. An unpredictable and unstable policy climate is counter-productive to attempts to encourage business to become more environmentally responsible — to quote the BCA report:

> As things stand now, our environmental policy may be a bigger mess than our environment. Market-based approaches that join the hard head of the accountant to the soft heart of the environmentalist offer the prospect of a genuine improvement; more clean-up for less money. It is an offer society should not refuse.

Market-based solutions to environmental problems come to terms with real costs to users. Society's environmental assets 'do not come with a price tag on them' according to the BCA, but economists have proposed certain factors to consider when determining what these assets cost the community, including the following:

- the value of production forgone because of environmental damage
- the value of earnings lost through pollution-induced health problems
- the value (price) people pay to prevent environmental damage
- the costs of repairing environmental damage
- how much people are willing to pay to visit or ensure preservation of a particular environment.

The BCA sums up its proposed instruments of environmental management and the benefits of each in Table 4.1.

Table 4.1 Components of a market-based approach to environmental management

Instrument	*Desirable Feature*
Conferral of private ownership rights on some publicly owned land	Conditions of public access to be clearly stated
Correct pricing of natural resources	Price to reflect social value of natural resource
Tradable emissions rights	Overall allowable pollution level to reflect value of environmental damage
Product charges for environmentally harmful products	Charges should reflect potential environmental impact
Charge for effluent and waste disposal	Charge should reflect environmental use (damage)
Enforcement incentives: • non-compliance fees • rehabilitation bonds	Levels to be set according to potential damage

Source: Business Council of Australia, 1990

The BCA maintains that government should re-examine incentives for industry to engage in applied research into environmental control such as recycling and emission control. However, caution is required to ensure that government agencies do not promote their own interests by exaggerating the public's perception of crisis. The BCA emphasises that attitudes to nature, and confidence in technology, vary markedly at different times and may 'reflect social moods which don't appear to have any rational basis — it is easier to alarm and scare people than it is to 'unscare' them'.

4.5 Environmental and capital wealth

Pearce and his co-workers (1989) tell us that our simplistic concept of wealth as only capital wealth requires an overhaul if sustainability is to come within our reach. They suggest that we should distinguish between two types of wealth:

- capital wealth, which can be used to achieve intergenerational equity by leaving to succeeding generations at least as much capital as the current generation inherited;
- environmental wealth, which can be used in the same way as capital wealth to compensate future generations.

Given these two legitimate forms of wealth, what do the trends in agricultural exports and land degradation tell us about the wealth of Australian rural areas? The increasing levels of degradation have already been referred to and there is mounting evidence to support the conclusion that as a nation Australia has been living off its land capital; that is, our capital wealth has been gained at the expense of our environmental wealth. In view of this general trend, the changes in the rural sector's contribution to national productivity, employment and exports are particularly important. The trends are shown in Table 4.2.

Table 4.2 Contribution (%) of the Australian rural sector

	GDP	*Employment*	*Exports*
1956–7	14	10.8	80
1966–7	10	8.3	68
1976–7	5	6.2	46
1985–6	4	5.8	35
1989–90	4	5.0	24

Source: Australian Bureau of Agricultural and Resource Economics Annual Reports, 1988–1991

4.6 Matching sustainability to carrying capacity

The idea of land capability analysis gave a basis for determining which uses each portion of the landscape was suitable for. It may be said that western societies moved into the so-called environmental era when laws and regulations were set up to control land uses and pollution.

More and more planners are becoming engaged in reassessing

what the land's capacity (or capability) really means. Integral to capability is the idea of human carrying capacity — a concept proposed by Verhulst as early as 1838. In terms of ecosystems, carrying capacity is 'the maximum population that can be sustained in a habitat without the degradation of the life-support systems' (Odum and Franz, 1980). When applied to human systems, carrying capacity becomes a more complex concept because it includes both cultural and technological dimensions, not to mention the additional effect of trade. For instance, the global carrying capacity will be considerably lower (in fact about 70 times lower) if all humans aimed to live in a United States-type culture, as opposed to a Bangladesh-type culture. In addition, by allowing substitution (e.g. oil for fuelwood) and increasing efficiency (e.g. irrigation), technology greatly increases the carrying capacity of one region at the expense of another.

In applying the natural ecosystem analogy to human systems, we must extend our ecological concepts to include cultural influences, international trade and economic trade-offs in a way that properly accounts for consumption. The city-states of Hong Kong and Singapore are often quoted as extreme examples of how human systems affect carrying capacity.

Bryson (1989) of the University of Wisconsin has attempted to calculate the absolute limits of the earth's human carrying capacity. Assuming that crops can capture 4% of solar energy and that about 5% of biomass is edible, 0.09 hectares of land is required to feed one human or 5.5 people per hectare. There are presently 3 persons per hectare, but at current rates population will double in 41 years. Bryson therefore predicts that we shall reach our 'sunlight limit' in 35 years. Although many planners challenge this prediction, even the most optimistic estimates of global population indicate a total of 10 billion by 2010 (Choucri, 1991). If science could increase our sunlight capture to 6% and double the edible proportion of our biomass to 10%, Bryson's timeline would be extended by only 20 years.

When considering Australia's human carrying capacity we find a geographical situation different from any other in the world in terms of space available per capita and demographic concentration — nearly 90% in urban settings, virtually all within 100 km of the coast. A small organisation called Australians for an Ecologically Sustainable Population Inc. was founded in 1988 to extend public awareness of the limits of Australian population growth from the ecological and social viewpoint. The basic tenet of AESP is that the survival of an ecologically sustainable population depends in the long term on its renewable resource base, and it recommends near-replacement fertility levels and low immigration rates. AESP's first goal is to establish a federal population policy. Young (1989) is critical of Australia's method of determining policies for future population growth; she maintains that little attention is given to the environmental and resource constraints. In recent years Australia's population growth rate (currently 1.4%) has been the highest in the developed world. Immigration has varied from 50,000 to 150,000 (103,000 in 1993). With a continuation of the 1989 level of fertility (net reproduction rate of 0.9) a constant net migration rate of 50,000 per annum from, say, 1956 would be sufficient to keep the population increasing to the middle of the next century and then gradually reach a near-constant population of 20M. The same result would be obtained from a net reproduction rate of 0.8 and a constant net migration of 100,000 per annum or even by rates of 0.7 and 150,000 respectively (Young, 1989).

Some observers believe that Australia persists with the old 'populate or perish' policy that assumes that unless Australia keeps increasing its population, it will be taken over by outsiders — 'share your resources or they'll take them'. However, the population debate goes far beyond people numbers and includes quality of life as an important outcome. A simple way to express the relationship is: I = PAT where I is environmental impact, P is population, A is level of affluence (consumption) and T is the kind of technology (including pollution) used to produce the affluence. Given this relationship, it is surprising that environmental groups in Australia have largely avoided the local population debate. Moore's (1990) study shows that the Australian Conservation Foundation formerly took an international humanitarian stance based on anti-racism and social justice which concentrated on reducing consumption per capita, changing pollution technology and disposing of waste. However, at the annual general meeting of the Australian Conservation Foundation in December 1992 the foundation called on the Commonwealth Government to 'act urgently to stabilise Australia's population', but omitted to suggest a specific level of stability. The ecological constraints of this continent are too often overlooked by community leaders such as the Catholic Archbishop of Melbourne, who claimed in 1988: 'This is a country of great emptiness. We should promote the opening of our shores to people in need, especially refugees'. (Moore, 1990). Figure 4.1 shows that the 'great emptiness' is largely dry country unsuited to cropping, which suggests that the extent of the cropping zone should be the basis of population expansion planning.

It is not appropriate to analyse Australia's immigration policy further here, but the warning that this country should avoid what are called 'lifeboat ethics' warrants reconsideration. The analogy is that Australia is a lifeboat and if too many victims are taken on board they will reduce the Australian passengers' chances of survival — may even sink their boat. Thus the philosophical question arises: is self-preservation to be construed as selfish, racist, inhumane, nationalistic or isolationistic? Here we must leave this particular dilemma, but in passing, let us be mindful of the national Australian opinion polls on immigration over the past 40 years, as shown in Table 4.3.

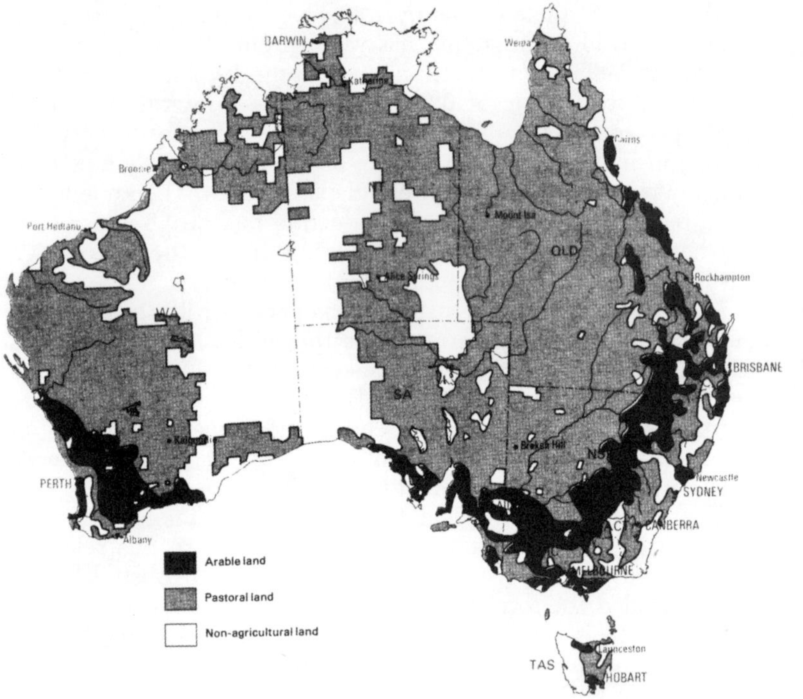

Figure 4.1 Arable and pastoral zones of Australia
WGSA 1991

Table 4.3 Attitudes to immigration (expressed as percentage of respondents)

Year	*Too high*	*About right*	*Too low*	*Don't know*
1961	16	37	43	4
1970	38	45	12	5
1981	45	37	11	7
1991	73	16	9	2

Source: Betts, 1993

Chapter 5

LAND USE PLANNING: THE BIG PICTURE

This chapter examines the national framework for a balanced land use policy and shows that in many regions agriculture is only one of several appropriate uses. We examine the competing forces behind land and water allocation in Australia's developing economy, show how the ecological limitations of different land forms act as constraints on intensive use, and review the effects of urban and rural development on the water quality of our rivers in the light of today's priorities.

5.1 Balanced land use policy — agriculture's framework

If agriculture is to be integrated into a broad land use plan in each state, the principles, goals and policies for national land use require clarification. Hallsworth (1977) wrote extensively on land use policies and suggested that in essence rural land use suffered three basic constraints:

- Wide climatic fluctuations, leading to great variations in yield.
- Unpredictable rises and falls in product prices in response to local supply and overseas markets.
- How the small size of many holdings limits the producers' options.

Besides the biophysical limitations imposed by nature on the capability of each portion of land, Hallsworth suggests four major determinants of land use for consideration in policy formulation:

- A large portion of Australia's export earnings is derived from extensive cropping and grazing.
- As energy becomes more expensive, lower energy-consumption systems (fuel and fertiliser) will be required.

- As population approaches 20M by the year 2000, a greater proportion of total food production will be required for domestic consumption.
- As allocated water reaches the limits of availability, competition between rural and urban demands will increase.

From these sets of points it is clear that planners must weigh up many other factors when choosing from land use options, and that a full range of options is not always available. Figure 5.1 represents the relations between resource management and planning.

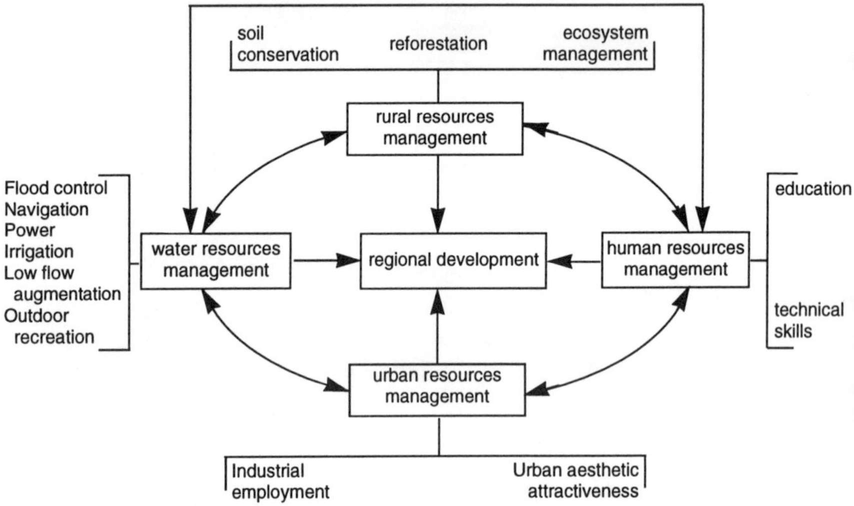

Figure 5.1 Relationships between resource management and regional planning
O'Riordan, 1971

In his attempt to put all the major land use goals in some kind of order, Cocks (1992) has developed fifteen land management goals, each with policy guidelines. His goals are in three groups: conservation, production and social, and while Cocks' concepts are not proposed in the form of a comprehensive national land use policy, they represent one of the most useful planning guides published to date. Here is a summary:

GROUP A: CONSERVATION GOALS

Goal 1 Maintenance of national soil productivity: Policy should concentrate expenditure on highly productive soils at risk, on areas where water supplies are threatened, where cost-effectiveness will be high and where landholder groups are positively motivated.

Goal 2 Maintenance of national water supply and air quality: Policy should concentrate expenditure on areas of high population, where pollution is likely to be irreversible and where investment in pollution control is cost-effective.

Goal 3 Maintenance of biodiversity: Policy should concentrate on areas of high diversity where cost-effectiveness is likely, and should minimise clearing and protect wetlands. Major exotic predators should be controlled and weeds of greatest commercial significance should receive priority.

Goal 4 Protection of historic and cultural sites: Policy should concentrate expenditure on valuable sites under threat, sacred and relic sites, sites subject to heavy use, sites properly evaluated, and where cost-effectiveness can be assured.

Goal 5 Development of a high-quality national park network: Policy should give priority to expenditure where parks do not have to be purchased, where threatened species occur, where diversity is wide, where species are not well represented in parks, where near-pristine conditions exist and where conservation is the parks' prime purpose. Policy should give priority to investment where management requirements are low, where no plans exist in heavily used parks, where ecosystems are threatened by exotics and where parks are close to cities.

GROUP B: PRODUCTION GOALS

Goal 6 Continued availability of primary production resources (farmland, forests, minerals and fishing): Policy should protect farmland from encroachment for urban uses, and minimise erosion, salinity, structure breakdown and acidification. Policy should establish hardwood plantations and protect native forests, and should allocate equitable and sustainable marine fishing quotas.

Goal 7 Socially beneficial resource development: Policy should invest in projects that have sound impact assessment, should account for all social and private costs and should support socially beneficial projects in spite of poor commercial prospects.

Goal 8 Maintenance of infrastructure to continue resource-based industries: Policy should maintain rural settlements and their services, and should minimise the socio-economic effects of reduced production in primary industry.

GROUP C: SOCIAL GOALS

Goal 9 Development of an effective national communication and transport network: Policy should give priority to eliminating freight delays and concentrate on areas where predicted demand is greatest. Policy should encourage energy-efficient transport in rural and urban areas.

Goal 10 Development of an effective network for the supply of water and energy: Policy should invest primarily in areas where supplies are substandard, where competition for water is high, where water quality is predicted to decline and where it can be improved at low cost. Policy should concentrate on localities where cheap water and power could stimulate the economy, and should give priority to cheap power generation by a range of alternative means.

Goal 11 Improvement of urban infrastructure and services: Policy should encourage the improvement of community services in deficient urban areas and in non-metropolitan locations that expect high population growth. Policy should give priority to urban situations where infrastructure can be cheaply upgraded and where deficiencies are most likely to cause social disruption. Policy should work toward keeping to a minimum the number of people requiring infrastructure.

Goal 12 Protection against natural and induced hazards: Policy should foster the protection of life and property in climatic hazard zones, and should minimise production losses due to climate.

Goal 13 Development of an effective network of recreation areas: Policy should invest primarily in natural areas which attract high numbers of visitors, are in easy reach of cities, and which offer special types of recreation. Policy should concentrate on easily acquired and easily managed areas.

Goal 14 Meeting the legitimate land needs of Aborigines: Policy should encourage the acquisition of lands which have special significance, where the Aborigines are particularly underprivileged and where claims can be substantiated. Policy should encourage resource-based Aboriginal enterprises.

Goal 15 Properly evaluate land use options where conflict occurs: Policy should encourage comprehensive investigation of the benefits of each option for areas of conflict and should pre-empt potential controversy by forward planning. Expenditure should be concentrated in areas of present and potential heavy demand by one or more alternative land use.

5.2 Planning for ecologically sustainable land use and agriculture

In 1991 the federal government established a task force for Ecologically Sustainable Development (ESD). Working groups were set up to report on each sector of the national economy, e.g. agriculture, mining, tourism, manufacturing, and to examine the extent and causes of (and the proposed actions to correct), any ecological unsustainability apparent in each industry. In coordinating the findings of each working group, the intersector report grouped its guiding principles for ESD into six spheres:

1 Global issues.
2 Economic diversity and resilience.
3 Risk management.
4 Biodiversity conservation.
5 Intergenerational equity.
6 Equity in this present generation.

Each sector contributed to these spheres but in the agricultural sector eight major areas of sustainability were identified: Land management, water use and quality, the use of chemicals, agroforestry, animal welfare, biodiversity, measuring sustainability and Aboriginal issues. Each major area can be divided into subsets, each with their symptoms, causes and corrective actions, as discussed in Chapters 8 and 9.

Australian agricultural systems are traditionally divided into two simple types — summer rainfall and winter rainfall systems. These are subdivided into cropping and grazing systems. Within these broad divisions, the degree of environmental hazards such as erosion, salinity and soil structural decline depends on soil type and local topography.

Williams and Hamblin (1991) have proposed eleven agro-ecological zones as a basis for resource management analysis. Their zones are primarily rainfall zones related to tropical and temperate temperature ranges (see Figure 5.2). Within these eleven zones a further 46 regions termed 'Agro-Ecological Regions' have been provisionally mapped to show where each form of land degradation ('resource management issue') occurs in each locality. Each map has a table of regions and issues that converts the geographical distributions to a matrix form showing which problems of unsustainability require attention in each region. Maps have also been drawn up from information supplied by each state, so that each major soil problem — e.g. water erosion, wind erosion, salinity, soil structure decline and acidity — can be plotted. These maps offer land use planners and managers Australia's first geographical database of degradation types, and make it possible to calculate total areas in need of corrective action and to estimate predicted costs for budgets.

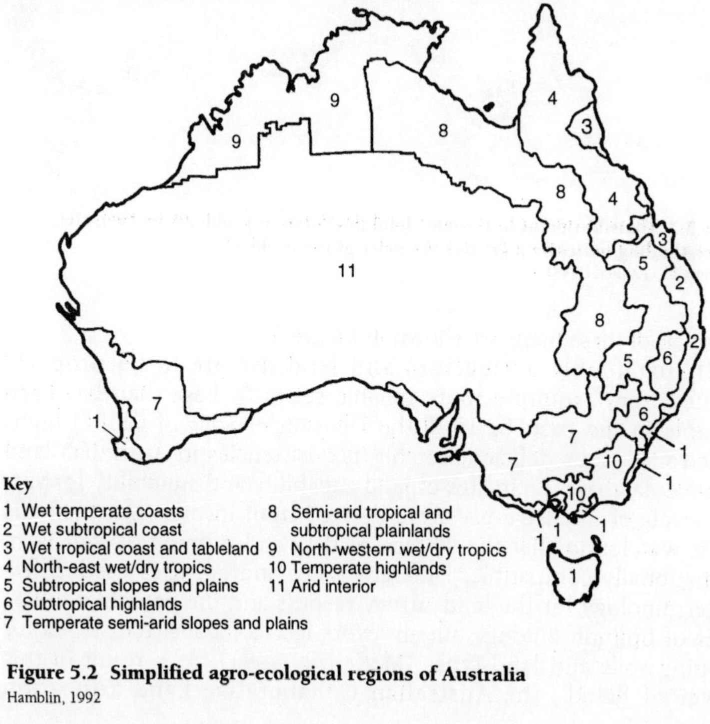

Figure 5.2 Simplified agro-ecological regions of Australia
Hamblin, 1992

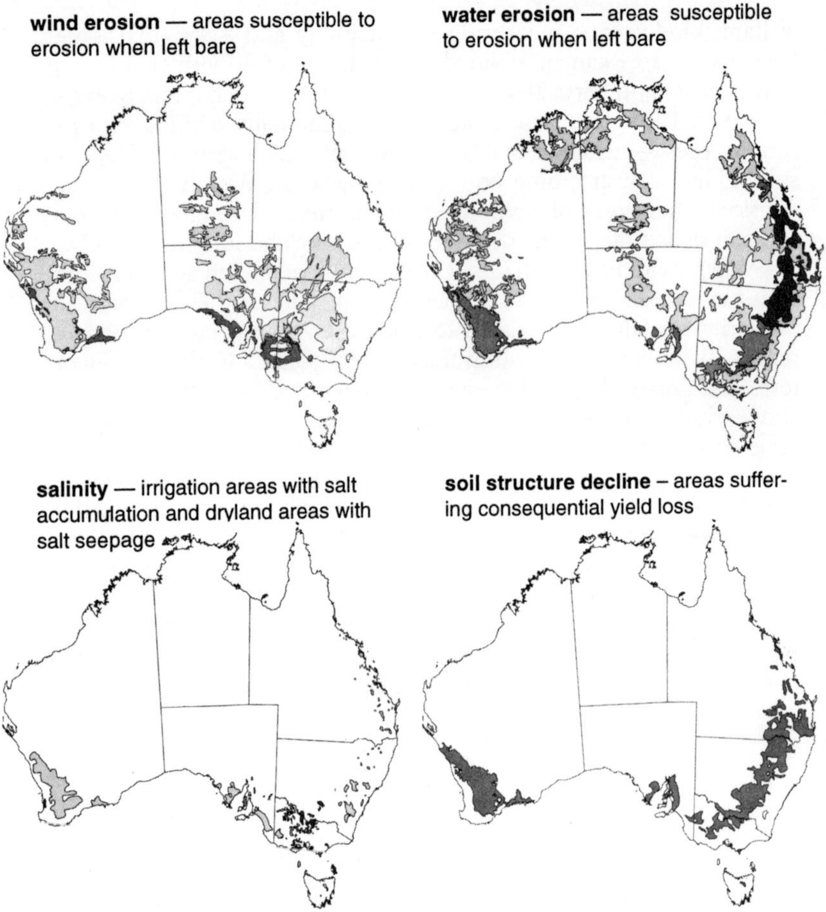

Figure 5.3 Distribution of four major land degradation problems in Australia. Darker shading indicates a greater intensity of the problem.
Commonwealth of Australia, 1991

Examples of these maps are shown in Figure 5.3.

If sustainable agriculture and land use are to be properly planned, they require a more reliable scientific base than has been available in the past. In 1990 the Division of Soils of CSIRO highlighted numerous deficiencies and inconsistencies in Australia's land resource database. Estimates of land capability and suitability for various levels of intensive use were derived from inconsistent methods and it was found that the storage of data and training of staff were not nationally compatible. Different states and agencies used different terminology for the land survey records and the perennial differences of opinion among soil surveyors led to inconsistent bases for mapping soils and land types (McKenzie, 1991). As a result of this 'Tower of Babel', the Australian Collaborative Land Education

Program (ACLEP) was established in 1992 to set standards, gain uniformity, and produce a comprehensive national system of land resource survey, classification and evaluation. ACLEP today is the collaborating body for land resource agencies and practitioners and is developing a network of Land Reference Sites and revising the Australian Soil and Land Survey Handbook.

The cost-benefit ratios of well-planned land resource surveys are very favourable when the results are applied appropriately to sustainable land use options. However, modern Australia requires much more than the earlier agricultural land use capability maps. We need to identify areas for waste disposal, areas of high susceptibility to erosion, salinity, acidity or structural breakdown, and areas of water recharge. Good land survey also provides a basis for planning urban foundations and industrial developments such as the optimal routing of underground cables. In addition, soil surveys serve as the best basis for planning the expansion of cities, cropping, forestry and rural residential estates.

5.3 Water allocation and quality control

Sustainable land use on the world's driest continent must include water with soil and vegetation as the major elements of the landscape. As a nation, Australia has always been aware of the survival value of water, but more particularly of the development value of water — the key to national growth. After land clearing, irrigation must rank as the second great strand of rural development. The Murray-Darling and the Murrumbidgee irrigation schemes, followed by the Snowy Mountain Scheme (also for hydroelectricity) are quoted as the great technological triumphs of this country. Little did the planners realise the threat which wetland salinity would pose to Australia's major irrigation schemes so early in the nation's history.

In earlier years the emphasis was on harnessing the wasted potential of great rivers flowing unused into the sea. In more recent times the emphasis has changed to water quality — quality affected by catchment management and pollution, but also affected by irrigation methods. Because large cities such as Adelaide depend on rivers like the Murray for their domestic water supplies, the economic implications of water quality go far beyond what is adequate for irrigation.The 1980s saw an upsurge of concern for an earlier idea, termed 'catchment planning', now called 'integrated catchment management'. The term implies an interdependence between all who live in a catchment, with special emphasis on the waste products that are disposed of or enter randomly into the river. As the national population increases, so two major trends have become important:

- The demand for water increases, leading to new pressures for competitive use and rational pricing of water for all uses.
- The accumulation of pollutants becomes more intense and

widespread and the types of toxic materials become more diverse.

In the early 1990s water quality conflicts occurred in several regions: severe salinity in the lower Murray, poisonous levels of blue-green algae in the Darling, and a significant rise in nutrients in the Barrier Reef waters from agriculture and towns. While these problems are being approached primarily as water quality problems, there

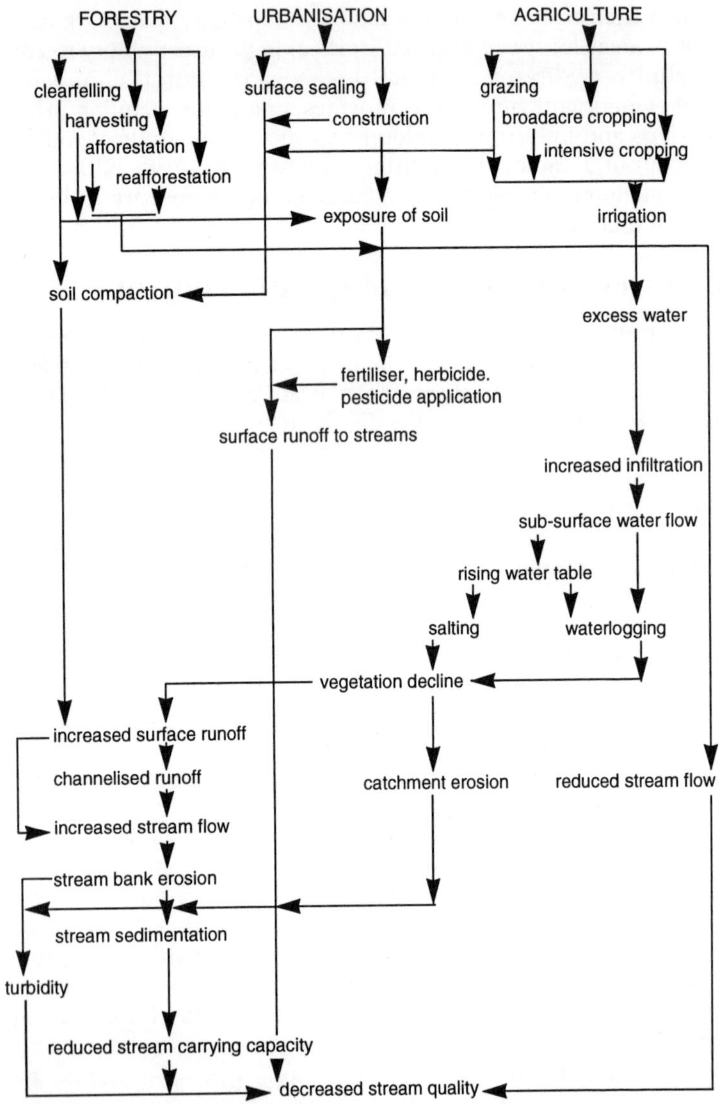

Figure 5.4 Relationship between land management and water quality

Castles, 1992

is no doubt that they are directly related to river flows and thus to water quantity and allocation issues as well. The complex relationships between alternative land uses and water quality are shown in Figure 5.4.

Today water management has become such a crucial element of sustainable land use that virtually all rural planning now aims for integrated catchment management. Part of the problem has been that in the past irrigation water was over-allocated so that many of the original ecological processes which depended on the periodic flushing of river systems have been degraded by lack of sufficient flows. It is likely that over-use of this limited resource will become a major sphere of conflict in the planning for the coming decade.

When all the environmental effects of modern intensive agriculture are compared, water pollution is probably the most damaging and widespread offsite effect of this industry. Non-point source pollution accounts for about half of all water pollution, and agriculture is the largest contributor to it. Note that the prime US research organisation (NRC, 1986) considers erosion's effects on water supplies to be greater than its potential effects on productivity, and the USDA (1987) estimates that the economic cost of off-farm water pollution is 2–8 times the value of erosion's effect on crop losses. It is also estimated that 50–70% of all nutrients reaching surface waters originate from agricultural land in the form of fertilisers and animal wastes. The N is carried in solution and the P is usually carried attached to sediment particles.

In recent years the wide range of synthetic herbicides used in agriculture has become cause for concern, partially because of the extent to which many common pesticides easily leach into groundwater (see Table 5.1).

Table 5.1 Relative mobility of 10 herbicides which are or were widely used in Australia

Herbicide	*Description*
Atrazine	Leacher
Bromacil	Leacher
2,4-D	Marginal leacher; heavy use
Dalapon	Leacher
DCPA	Leacher
Dicamba	Leacher
Diuron	Leacher
MCPA	Marginal leacher
Picloram	Leacher
2,4,5-T	Marginal leacher

Source: National Research Council, 1989

Because pesticides are more likely to affect human health than herbicides are, the significance of the entry of pesticides into groundwater used for human consumption is receiving increasing attention from medical authorities (see Table 5.2).

Table 5.2 Confirmed pesticide detections in groundwater due to normal agricultural use

Pesticide	*Health advisory level (parts per billion)*	*Median concentration (parts per billion)*
Aldicarb	10.0	9.00
Aldrin	10.0	0.10
Atrazine	3.0	0.50
Chlordane	0.03	1.70
1,2-D	0.0013	4.50
1,3-D	0.20	123.00
DCPA	3,500	109.00
DDT		1.70
Diazinon	0.63	162.00
Dieldrin	0.00219	0.02
Lindane	0.026	0.10
Methyl parathion	2.0	88.40
TDE	0.031	4.80

Source: National Research Council, 1989

Case studies reflecting water quality problems are everywhere to be found, especially but not only in the Australian coastal zone. Water pollution originates from a wide range of land activities, but for the purpose of case studies three major types of water quality problems have been selected: farm runoff, salinity and heavy metal pollution.

CASE 1: AGRICULTURAL RUNOFF IN THE HAWKESBURY-NEPEAN CATCHMENT OF NEW SOUTH WALES

This system delivers both agricultural and urban runoff, combining the sediment, fertiliser and pesticides from farming with the town wastewater and sewage from Sydney's outlying areas. The national water quality guideline for total P is 10–100 μg/L (μg = microgram or 10^{-6}g). In the Hawkesbury-Nepean the total P content in 1992 was in the 12–160 μg/L range, which is well within the range in which algal blooms occur. National guidelines for N are 100–750 μg/L while this catchment measured 1,000–2,750 μg/L. As expected, analyses showed that nutrient concentrations increased as flow rates decreased, concentrating the pollutants. The P output from feedlots is becoming a major source of nutrients in Australian rivers. The P cycle from this source is shown in Figure 5.5.

CASE 2: HEAVY METAL FROM INDUSTRY IN THE DERWENT RIVER OF TASMANIA

A major zinc factory on the upper Derwent has been identified as the source of much of the heavy metal entering the Derwent above Hobart. Water testing has revealed cadmium, lead, mercury, copper and zinc. In this case the water quality happens to meet the requirements for drinking water but is above the limits required for the protection of aquatic ecosystems (including seafood production systems).

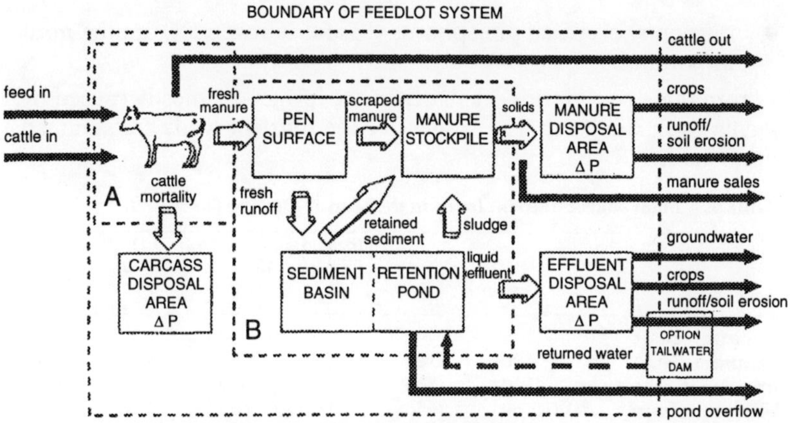

Figure 5.5 Flow paths of phosphorus in a beef feedlot system
Crone, 1990

As an example, lead limits are set at 1–5 μg/L for ecological protection, but the Derwent, downstream from the zinc works, carried 6.4 μg/L in 1986. Mercury limits are set at 0.1 μg/L but the river currently carries 0.2 μg/L, so the marine ecosystems may well be at risk.

CASE 3: SALINITY IN THE HUNTER RIVER OF NEW SOUTH WALES

Acceptable quality water should have a salinity range of 280–1,500 μS/cm (μS = microsiemens, a measure of electrical conductivity related to salt content). At various locations along the Hunter, the river water registered 350–1,200 μS/cm in 1992/93. Although a range of salt levels is satisfactory for various water uses, the NSW Environment Protection Agency (EPA) sets the limit at 700 μS/cm for acceptable water quality (ANZECC, 1992). Unlike many situations in the Murray-Darling Basin where tree clearing and inherently saline soils are the major causes of salinity in river water, in the Hunter Valley mining and low flows are the major causes. Wastewater from opencut coal mines in geological formations with a high salt content has a marked effect on river water salinity. In addition, the large powerstations discharge cooling water after it has evaporated to a more concentrated salt content. The over-allocation of the irrigation capacity of the Hunter has removed much of the earlier flushing effect of higher flows which maintained lower salt levels.

The quality aspects of water supply in Australia are not only complex but are becoming more widely acknowledged as important elements of sustainable development, so it is worth examining closely what is reducing water quality. Returning to the Murray-Darling Basin, two groups of the quality criteria are considered:

- phytoplankton (algae), which include over 200 species in four groups: Diatoms, Green Algae, Flagellates and Blue-Green Algae;

- nutrients, of which phosphorus (P) and nitrogen (N) are the most important.

Within these two groups, increases in the former are mostly caused by increases in the latter. The sources of the nutrient loads are summed up in Table 5.3.

Table 5.3 Point source nutrient loads in the Murray-Darling Basin

	Nutrient loads (tonnes/year)					
	Dry Year		*Average Year*		*Wet Year*	
Point source	*t P*	*t N*	*t P*	*t N*	*t P*	*t N*
Municipal sewage Treatment plants	500	2780	500	2780	500	2780
Irrigation drainage	110	630	170	980	260	1480
Urban stormwater	60	450	80	660	120	1000
Total	670	3860	750	4420	880	5260

Source: Mackay, 1992

While increased nutrient content (known as eutrophication) is recognised as the prime cause of Murray-Darling algal blooms, such algae are also increased by reduced river flows (including over-allocation) and by degradation of the 'instream ecosystems'. The Murray-Darling Commission has proposed a step-wise approach to problem-solving, as follows:

1 Provide an understanding of the physical/biological processes.
2 List control options.
3 Analyse the costs and benefits of each option.
4 Recommend appropriate action.
5 Propose further research priorities.
6 Select monitoring mechanisms to evaluate and modify controls where necessary.

A three-stage strategy has now been proposed to control the Basin's water quality. It is agreed that in the short-term the river flows must be manipulated and the water in the storages must be destratified (mixed). For the medium term, nutrient sources must be reduced by decreasing point source discharges of P and N and herbicides, while at the same time increasing river flows by greater releases from storages. Finally, long-term solutions based on improved catchment management, rehabilitating streambanks and riparian vegetation and preventing livestock access to streams are recommended as the basis for the permanent stability of the river system (Mackay, 1992). This long-term management will be based on a markedly changed vegetation and land use in the Basin as shown by the comparison of the original and present Basin situation in Table 5.4.

Table 5.4 Changes in areas under different vegetation types in the Murray-Darling Basin from the 1770s to 1980s

Vegetation	*Forest & open forest*	*Woodland*	*Open woodland*	*Tall shrubland (mallee)*	*Low shrubland (saltbush)*	*Natural grassland crops & pasture*
Area of each in the 1770s	162	533	143	124	64	40
Area of each in the 1980s	58	199	324	82	54	349
% Change in area of each	-64%	-63%	+127%	-34%	-16%	+773%

Source: Walker, Hatton & Mundy, 1992

Chapter 6

LAND USE PLANNING AND AGRICULTURE

Having considered which factors should dictate land and water use in the previous chapter, we now examine in some detail the conflicts and objectives of rural land use planning. This chapter gives special attention to the protection of prime agricultural land against urban expansion, and explains the major activity of property management planning in the context of land use suitability classes. We conclude with an examination of the present state of knowledge on predicted climatic changes and how we might prepare for such changes.

6.1 Goals and conflicts in rural land planning

Within the ambit of Ecologically Sustainable Development (ESD) the planning and management of rural lands forms a major subset. Almost invariably, when rural lands close to cities or national parks are available for alternative uses, a conflict of interests arises. Capelin (1993) has given a useful overview of land use planning in Queensland and suggests the following as aims for planning in districts suited to agriculture:

1. The protection of resources suitable for agriculture.
2. Making available properties large enough to be economically viable under agricultural production.
3. The encouragement of economically and ecologically sustainable management practices.
4. The protection of fauna and flora resources.
5. The protection of areas adjoining agricultural land from the downstream or drift effects of agricultural practices (chemicals, sediments, odours, noise, dust).
6. The resolution of conflicts between adjoining land uses.

While this list may appear straightforward, in practice there are considerable problems with reaching agreement on land use priorities, and with resolving conflict between neighbours. The extent to which good agricultural land should be preserved for food production depends on whether economic forces make it unacceptable to leave it to the marketplace to rezone such land for residential and other purposes. Almost invariably cropland will fetch a 400–500% higher price for subdivision than for farming and the pressure to subdivide is increasing every year. Capelin (1993) gives the example of the expected population increase of 450,000 in southeastern Queensland in the decade up to the year 2000, requiring an additional 25,000 ha of land. This means 2,500 ha of agricultural land would be lost annually. In the same way, the area under sugar cane production was reduced by 3,882 ha in the decade ending in 1990.

Integral to the aims of rural land use planning are procedures that reduce the fragmentation of agricultural land, not only in terms of maintaining sufficient area for the type of crop production suited to the region, but in terms of preventing the patchwork of alternative land uses which aggravates the opportunity for boundary disputes over uses. In earlier years the number of peri-urban dwellers on small subdivisions was not a significant factor in rural planning, but today the expansion of the '5 acre block' rural retreat for city dwellers has become a political problem. Local authorities continuously consider applications for subdivision, in many cases such selling-off of agricultural land is regarded as the landholder's equivalent of a superannuation entitlement. The remaining adjacent farmers are then faced with complaints about pesticides, herbicides and noise affecting the newcomers, while the farmers have to put up with dogs, rubbish dumping, theft, vandalism and new fire risks.

Besides the 'cocky/blockie' conflicts of interest, the allocation of rural land for national parks and tourism should also consider community needs in the broadest perspective. One of the most difficult issues facing planners of rural land use is the geographical extent of the 'stakeholders': to what extent should local residents' wishes be allowed to override the amenity value to the wider community?

The relative importance of each sector of the Australian economy varies from year to year, depending primarily on weather and world commodity prices. In recent years the ranking has remained reasonably constant, as shown in Table 6.1.

Table 6.1 Sources of foreign revenue 1988/89

	$ billion
Mineral products	19.0
Farm products	15.5
Tourism	6.5
Forest products	0.5

Source: Department of the Arts, Sport, the Environment, Tourism and Territories, 1989

More current trends in the ratio between minerals and farm products are shown elsewhere, but by and large the allocation of land should reflect the relative importance of these wealth-generating enterprises. It is predicted that tourism, particularly eco-tourism, will increase its market share significantly by the year 2000.

6.2 Zoning as a basis for rural land allocation

In the solution of conflicts over rural land use planning, the idea of allocating all land within certain defined boundaries to one prime purpose has become known as *zoning*. In this way the local authority, usually the shire council or some other local government planning office, designates mapped areas to each of several alternative uses: e.g. rural residential, agriculture, grazing and timber production, mining, national parks, and so on. Such allocation of portions of land through the zoning process is usually based on the capability of the land and on the development needs of the region or shire. In theory at least, the zoning of land (a) reduces conflict between incompatible land uses, (b) uses the land most appropriately according to its potential and its hazards, and (c) offers a practical framework on which landholders can plan and operate their enterprises.

In one sense, zoning may be regarded as a growth-controlling mechanism that gives some certainty to purchasers and lessees. However, in some categories or zones applications for rezoning (change in land use) are allowed every five years or more. This means that land presently zoned 'agricultural' may be reconsidered for rezoning to 'rural residential' for instance, twice each decade. The public may believe that the zoning procedure guarantees good planning, but this depends on the value-judgements made on what constitutes 'appropriate' uses for each of the zoned areas.

Zoning was originally used only in urban areas to separate, for instance, the industrial from residential sectors of a city. When zoning is applied to rural areas, the shortcomings and dilemmas of the zoning process become more obvious and more complex. In urban settings, zoning had its origins in environmental quality control (noise, air pollution, heavy transport) but in peri-urban and rural settings the issues are less clear-cut and much more subject to outside influences such as state planning priorities or the need for more national parks.

Walther's (1986) analysis of the practical application of the zoning concept to rural lands or 'natural resource lands' indicates that despite the potential advantages of clarity, certainty and conflict reduction, zoning as a planning procedure has serious weaknesses. These are described in terms of 'five dilemmas of zoning in the management of natural resource lands':

- The *generalisation or classification* dilemma, which requires each portion of land to be grouped or classified into one of a predetermined set of classes or uses. In the process, much generalising and approximation is required to fit the inflexible class system, even when it is obvious that some internal sections of the mapped land zone are in fact better suited to other uses or zones. This is fundamentally a problem of artificially homogenising a diverse landscape.

- The *present land use* dilemma, in which the current uses and developments constrain easy change to an alternative, albeit more appropriate, use from a land capability viewpoint. The problem becomes one of attempting to impose a new order on an apparent state of disorder. Both the expense and social upheaval often leave this dilemma unresolved in societies that accept development much more readily than undevelopment. The result is that current land use often makes maps of proposed land use seem patchy, fragmented and not well correlated with inherent land capability.
- The *justification or rationality* dilemma, which arises from the somewhat vague criteria used in rural planning. Compared to the more structured criteria used in urban settings, decision-making on rural zoning is much more value-laden, even philosophical. The justification for rural land zoning assumes a greater role for forward-thinking and the provision of amenities, and its rationale can seldom escape the criticism of being no more than personal opinion or even emotional value judgement.
- The *problem abstraction* dilemma which assumes that land use conflicts can be solved by territorial space allocation. The provision of infrastructure such as access roads often acts as a practical constraint on the zoning procedure. In other cases the aesthetic values of rural landscapes cannot be measured or gauged by the monetary values applied to other more productive land attributes. The conflicts in rural land use also have a temporal (time) dimension as well as a spatial (place) dimension which makes the compatibility of uses even more complex.
- The dilemma of *uncertainty* — the insecurities that planning for a changeable future is fraught with. Land zoning must be based on present land use technology within the confines of present social priorities and goals. For instance, as Australian society moves from the pioneering era into the early-maturity of nationhood, so the primacy of production declines in favour of a rising value on amenities, natural beauty and environmental quality. While the democratic process can influence rezoning, the ultimate decisions are made by politicians, usually on a three-year term of office — a situation which encourages expediency at the expense of vision.

Despite all these difficulties, serious attempts by local authorities to apply zoning procedures in an equitable and flexible manner will probably form the basis for rural planning for many years to come. In Australia, the question is not whether zoning should be applied but *how* it should be applied.

6.3 Protecting good agricultural land by strategic planning

Few trends in rural Australia are more contentious than the loss of high-quality cropland to residential subdivision. The level of contention increases where particular agricultural sectors such as sugar production depend on a minimum area to justify their infrastructure and plant. In the case of cane production, the mill and its tramways represent a substantial investment for the growers, which in turn requires a minimum cane tonnage for viability.

Apart from industry viability, the fragmentation of agricultural land and the uncontrolled expansion of unserviced 'ribbon development' (single blocks adjacent to roads), rural subdivisions usually increase conflict between neighbours. The local authorities usually have discretionary powers to make exceptions to their own by-laws on subdivision. For instance, 20 ha is a common minimum size for rural allotments in areas zoned agricultural or rural, but shire councils regularly allow subdivision for additional family residences. Such subdivided titles then pass onto the open market and uncontrolled rural residential fragmentation only too often results. Capelin (1993) gives examples from the densely populated coastal shire of Maroochy in which over 66% of the rural subdivisions approved between 1988 and 1990 were discretionary approvals, i.e. exceptions to the rule. In the Redland Shire (south Brisbane) only 4.4% of good quality agricultural land remains in allotments greater than 12 ha, in a shire where 16–20 ha is the minimum allowable size for agriculture.

In an effort to promote orderly shire development that acknowledges both the potential of the land and the needs of various community sectors, many local authorities have adopted strategic plans to guide land use and protect the best agricultural land in perpetuity. Under strategic planning, the distribution of land among the major alternative uses is planned ahead at least for the coming decade, and a series of objectives for each land use or for each zone is included. Within designated locations of the strategically planned area, more detailed plans, termed Development Control Plans (DCP), can be made and implemented. In rural areas DCPs are appropriate only in special circumstances such as where subdivision or erosion hazard on steep escarpments cause special concern. DCPs can assist the application of sustainable agriculture on the basis of a legal framework which provides some degree of certainty and reduces conflict.

One of the problems with the strategic planning of rural land is the deficiency of reliable information on the land resources and their potential. While many areas have been mapped in some detail, there are still vast areas where no soil surveys and no topographic data have been mapped. Another problem is the continual pressure from developers, with or without encouragement from shire councils who benefit from subdivision, to convert good agricultural land to residential or industrial blocks. These losses of productive cropland are likely to continue if local authorities fail in their responsibility to preserve agricultural land, and leave land use trends to the marketplace.

The conflict over the intrusion of new communities into traditional agricultural areas is complex and laden with bias and subjective judgements. The results of subdivision in intensive agricultural regions extend beyond the serious loss of agricultural productivity —

newcomers complain that farming operations are incompatible with their (superimposed) lifestyle. In the case of Queensland cane farming, such complaints are made about pesticides, cane burning, and round-the-clock transport noise and dust for example, as shown in Table 6.2.

Table 6.2 Common causes of conflict between agriculture and urban interests in the north coast region of NSW, Australia

Conflict	*Explanation*
Absence	Farmer may have to assume more responsibility for bush fires, straying stock, trespassers etc while neighbour is at work or away.
Access	Traditional 'gentlemen's agreements' for access between farms may break down with the arrival of new settlers.
Catchment management	Design, funding and land, water and vegetation management plans complicated by larger numbers of landholders.
Cooperation	Absence, inability or unwillingness may prevent work sharing on-farm or in the community.
Dust	Generated by farm operations including cultivating, fallow (bare) ground, farm vehicles, livestock yards, feed milling, fertiliser spreading, etc.
Drainage	Drainage canals blocked from lack of maintenance or failure to cooperate with voluntary drainage authority.
Electric fences	Electric shocks to children, horses, dogs.
Fencing	Disagreement on maintenance, replacement, design and cost agreements.
Firearms	Shooting near livestock and homes. Disturbance, maiming and killing of livestock.
Flies	Spread from animal enclosures.
Litter	Injury and poisoning of livestock via wind-blown and thrown articles, also damage to farm equipment machinery
Noise	Use of farm machinery at night or weekends. Low-flying agricultural aircraft. Livestock weaning and feeding.
Odours	Odours from piggeries, feedlots, dairies, poultry, sprays, fertiliser, manure.
Pesticides	Their use, storage and disposal. Aerial spraying.
Pollution	Underground and surface waters contaminated with effluent, chemicals and pesticides.
Poisoning	Spray drift onto non-target plants. Poisonous plants, containers and effluent consumed by livestock.
Roads	Cost and standards of maintenance. Slow/wide farm machinery. Livestock droving and open front gardens.
Straying livestock	Fence damage, spread of disease and parasites.
Smoke	Disposal of crop residue, bedding, scrub and pasture.
Theft	Crops, livestock, fodder, machinery, equipment fencing.
Trees	Clearing for agriculture. Felling for sale.
Trespass	Recreational, sporting and vandalism.
Vandalism	Fences, machinery, buildings, vehicles, crops, livestock.
Water	Quantity, quality, legal rights and priorities.
Waste disposal	Effluent, garbage, spoiled crop, animal carcases,manure.
Weeds	Failure to control, spread to clean neighbouring land.

Source: Dent, 1982

As an example of local planning, when appropriate locations and planning controls for rural residential development are determined in New South Wales, the following criteria are considered (NSW Planning and Environment Commission, 1979):

1 The real demand for rural settlement allotments in relation to the capacity of adjoining urban areas (or the rural community development itself) to provide employment.
2 Rural settlement should not be permitted adjacent to urban centres in the path of likely urban expansion, because of the adverse effects and additional costs involved in later urban development.
3 Rural settlement should not be permitted in areas proposed as service corridors nor with direct access along main road systems, to prevent ribbon development and reduced rural amenity.
4 Consultation with the Department of Mineral Resources is desirable, to protect the potential extractive resources such as sand, gravel, brick-making clays and other minerals.
5 Consultation with the National Parks and Wildlife Service and the National Trust over designated areas or areas of significance for conservation.
6 Consideration of areas affected by the State's Coastal Protection Policy.
7 Areas of prime crop and pasture land to be avoided.
8 Remote areas with poor access and services to be avoided.
9 Areas of afforestation potential to be considered in rural community titles or similar developments.
10 Water catchment areas to be avoided.
11 Severe bush fire hazard areas to be avoided.
12 Natural beauty and scenic areas to be avoided.
13 Areas subject to flooding and coastal erosion to be avoided.
14 Areas susceptible to soil erosion to be avoided.
15 Areas of steep and unstable land to be avoided.
16 Areas costly to service to be avoided.

After considering the above criteria, land identified as suitable for closer rural settlement or rural residential development, including rural community titles or similar developments, should then be given a priority according to whether it is reasonably close to towns and villages with good access and where services are not unduly expensive and are readily available.

One of the most intense conflicts between agriculture and urban development is the sugar cane case study in Queensland. Sugar land is being lost to urban expansion and, more particularly, to rural residential (hobby farm) use. Some of this land is 6–8 times more productive than the average land for the state overall. In the case of urban expansion, such land is 'sterilised' and lost forever to this terminal use. When this happens, the productivity of the soil concerned is written off as a national asset. Coleman (1992) has reviewed the Queensland cane land situation as follows.

Australia has less than 10% of good arable land. Queensland has 2%, of which 1.6% is currently cultivated. The sugar industry has lost 1,800 hectares to other land uses in the past decade. Hambledon Mill near Cairns was closed as a result of this loss. In the Moreton region 600 ha of cane land has been lost, which is equivalent to an industry loss of $2M (at 11 t/ha and $320/t). The strategic plans drawn up by shires on the cane belt are not sufficient to protect good agricultural land. The gross value of sugar makes up 65% of the total gross agricultural production in the 26 shires in which sugar is grown in Queensland. This position is particularly important considering that the so-called 'multipliers' from sugar are 3.45 for income and 2.87 for employment, compared to 2.72 and 2.15 respectively for all agricultural production. For this reason the nation's agricultural resources cannot continue to be used 'as forms of social welfare or personal superannuation schemes' (Coleman, 1992). The gains of a few will lead to substantial losses by many other landholders and employees and it has been estimated that losses of less than 10% of present assigned cane land could lead to the closure of small mills in a single mill district. There are pressing reasons why rezoning decisions on agricultural land should not be left to local authorities in situations where the personal interests of both the shire councillors and elderly farmers work against sound long-term planning.

The sugar industry in this region will have to accept that its proposed expansion must meet zoning constraints on land suitability if it is to avoid the disastrous erosion that has been plaguing the northern sugar lands of Queensland, where whole hillsides were on the move with annual losses of over 300 tonnes/ha before green cane harvesting was introduced. The industry will have to take responsibility for environmental control and will have to accept that in future certain zoning decisions will go against the wishes of those who aim to expand the sugar lands. This will all be part and parcel of 'live and let live' in the future.

6.4 Classes of agricultural land

Since the 1940s the American and British farm planning procedures have been based on systems of land classification which match the intensity of production to the potential of each portion of land. Some systems had eight classes, others six, and the most recent have only four or five. An example of a recently adopted system proposed for Queensland is given in Table 6.3.

These classes indicate the degree of protection or precautions required on each class of land. As the classes become steeper and the soil thus shallower, so the amount of conservation planning and management increases. For instance level, deep soil may require only stubble mulching in a strip crop (alternate summer and winter crops) layout, with no earthworks or special design features. On the other hand, land with a slope of 3–6% usually requires contour banks and associated grassed waterways to convey water safely from the ends of the banks to the natural waterway. Land in the same area with a slope of 7–10% may require a combination of more closely spaced banks and stubble mulching, or annual cropping may have to be discontinued and the land used only for perennial crops such as pastures, perennial hay crops or even trees. In this way the old adage of 'using each portion according to its productivity and protecting it according to its hazards' is put into practical application.

Table 6.3 Classes of agricultural land

Class	*Land capability*
A	**Crop land** — suitable for current and potential crops with production limitations ranging from none to moderate levels.
B	**Limited crop land** — marginal for current and potential crops due to severe limitations, and suitable for pastures. Engineering and/or agronomic improvements may be required before it is considered suitable for cropping.
C	**Pasture land** — suitable only for improved or native pasture due to limitations which preclude continuous cultivation for crop production; but some areas may tolerate a short period of ground disturbance for pasture establishment.
D	**Non-agricultural land** — not suitable for agricultural uses due to extreme limitations. This may be undisturbed land with significant habitat, conservation and/or catchment values or land with very steep slopes, shallow soils, rock outcrop or poor drainage.

Source: Capelin, 1993

6.5 Farm planning — the ultimate planning unit

At a very local level, i.e. at the individual property level, the principles of sound rural land use planning are applied in the *farm plan*, or what today has become known as *Property Management Planning* (PMP). This procedure represents the ultimate local unit of land use planning and it is at this level that the most useful changes in management occur. It is thus also at this level that practical results and meaningful improvement in land condition can be expected.

Since World War II, farm planning has been an integral part of soil conservation extension in Australia and other countries influenced by the British agricultural advisory systems. The first step in farm planning has always been to map the natural resources within the farm boundaries — i.e. soil types, vegetation types (for grazing and timber), waterways and runoff flow patterns. From these physical features it is determined where the present and potential cultivated land lies and how excess runoff water can best be handled with minimum erosion and flood losses. The farm plan map shows which steeper land should be maintained under a dense grass cover for grazing, and indicates the watering points and fences planned to allow the most efficient grazing and spelling of all paddocks. On the cultivated land the positions of stormwater drains, contour banks and grassed waterways are shown on a mosaic (composite of aerial photographs) on a scale of approximately 1:12 000 as the base map. The soil types, flow patterns, fences etc. can all be superimposed on the mosaic with transparent overlays, one at a time or in combinations. Special erasible pens are used to make changes as the planning process proceeds.

In simple terms the farm plan has always aimed to provide a basis for using each portion of land according to its capability. Earlier plans in Australia concentrated on the soil conservation aspects; today's plans generally have three distinct but interdependent segments:

- The *physical planning* of the sustainable use of the farm's natural resources.
- The *management planning* of the farming operations concerned with crop and animal production.
- The *financial planning* of the farm business, which takes account of the economic sustainability of the farm's enterprises.

These three elements explain the term Property Management Planning, which today refers to the fundamental unit or building-block of the national Landcare Movement. Thus PMP is to sustainable agriculture what fuel is to an advancing army — it is the basic element of the national strategic plan. The management and economic analysis of sustainable agriculture on individual farms are dealt with in Chapter 10.

Table 6.4 CSIRO climatic predictions for 2030

Climate	*Changes*
Temperature	Up by: 1 to 2°C in northern coastal areas 1 to 3°C in southern coastal areas 2 to 4°C inland More in drier areas, less in the wetter areas
Rainfall	Large area average increase of 0–20% Less certain overall decrease of 0–20% in winter rainfall region General increase in rainfall intensities Possible marked increase in heavy rain events Longer dry spells in mid-latitude
Evaporation	5 to 15% increase in potential evaporation by 2030
Tropical cyclones	Could travel further south Effect on intensities still uncertain Preferred paths may alter Frequency change
Snow line	Up approx. 100 m per 1°C local warming
Winds	Weaker trade winds, but change not yet clear Stronger monsoon westerlies in northern Australia Mid-latitude westerlies further south Strong winds will occur with severe weather
Sea level	Global average rise of 3 to 10 cm per decade Best estimate for Australia by 2030 is about 20±10 cm above 1990 levels

Source: CSIRO, 1991

6.6 Anticipating climatic change

Much research has been done over the past decade into the so-called Greenhouse Effect. Henderson-Sellers and Blong (1989) have given a comprehensive explanation of what scientists believe is happening to our climate as a result of air pollution. Two questions persist, however: Is there really enough evidence to predict the size and direction of these changes, and how will the changes affect regional Australia? The CSIRO sums up its 'best bet' for the year 2030 in Table 6.4.

With predicted increases in summer rainfall and in intensity of rainfall, the effect of climatic change on cropping systems and soil erosion could be serious. While reliable data is not readily available, attempts to simulate runoff and erosion suggest that both these factors may more than double. Thus while the full effects of such a greenhouse scenario cannot be validated at this early stage, the possible outcomes must be taken note of by planners and researchers. In Table 6.5 McKeon (1988) and his colleagues present some possibilities for the black soils of Queensland's Darling Downs region.

Table 6.5 Component and combined effects of the greenhouse scenario on wheat production, run-off and soil loss on the Darling Downs

	Crop yield kg/ha/year	*Run-off mm/year*	*Soil loss tonne/ha/year*
Model calibration (1976–83)			
observed	2,805	59	44
simulated	3,033	68	49
Simulation (1912–84)			
present climate	2,868	45	32
temperature (+2°C)	2,684	46	33
winter rainfall (-20%)	2,650	40	29
summer rainfall (+30%)	4,070	135	96
winter and summer rainfall change only	3,863	129	92
full greenhouse scenario	3,532	132	95

Source: McKeon et al., 1988

With cyclonic influences predicted to extend up to 300 km further south than at present, and with greater rainfall generally over the top half of the continent, the effects of climate change may also be significant for the pastoral country. Carrying capacity, the frequency of fires and the encroachment of woody weeds could all increase under the predicted new regime. Graetz (1988) and his co-workers have given special attention to the possible shifts in the outer boundaries of the cropping zone, as affected by increases and decreases in rainfall in different regions. Their map (see Figure 6.1) requires more data before the trends it depicts can be substantiated, nevertheless these shifts must be borne in mind when assessing future trends in pasture condition and carrying capacity. They must also be taken into account when planning marginal cropland in several states, especially where a decrease in winter rainfall may necessitate a retraction of the present cropping margins in Western Australia and elsewhere.

It is difficult to anticipate climatic change in Australia because the evidence of change on regional climate, and thus on regional agricultural production, is not fully confirmed. Scientists are generally agreed that increases in greenhouse gases will cause general global warming, although a few such as Daly (1989) still insist that the present warming is a temporary change in a continual global cooling process and that scientific fraud and modern myths are being created by researchers eager for more funding. In particular our knowledge of the interactions between ocean and atmospheric temperature changes is deficient, as is our understanding of the timescales involved, i.e. the rate at which air temperature change affects sea temperature change. This air/sea relationship is of potentially great significance to Australian agriculture and pastoralism for if changes occur more rapidly than our current inadequate predictions indicate, the so-called El Nino/ Southern Oscillation system could cause severe, widespread drought (El Nino is based on sea temperature in the Pacific and Southern Oscillation is based on atmospheric pressure to Australia's north). Another unknown is the effect of 'sudden' (unpredicted) changes in ocean circulation, increasing

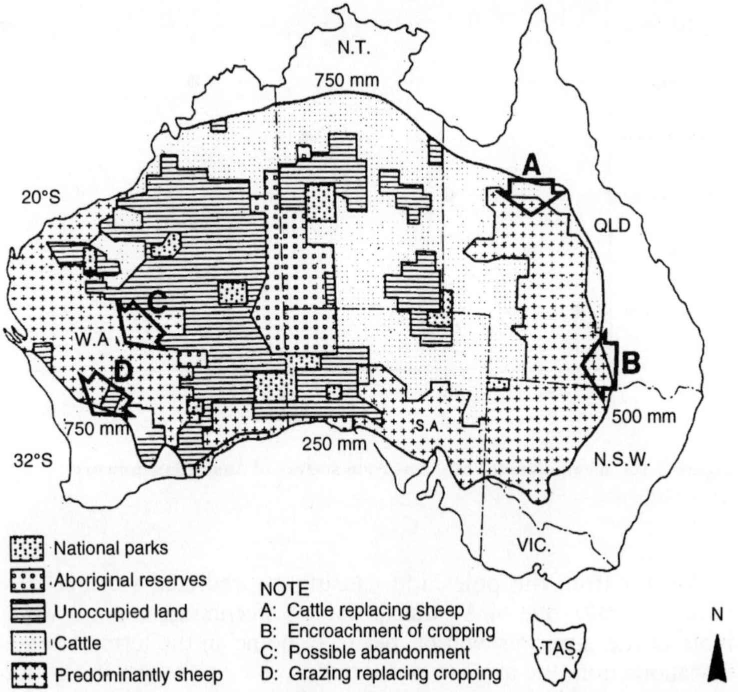

Figure 6.1 Predicted land use changes resulting from a climatic change scenario
Castles, 1992

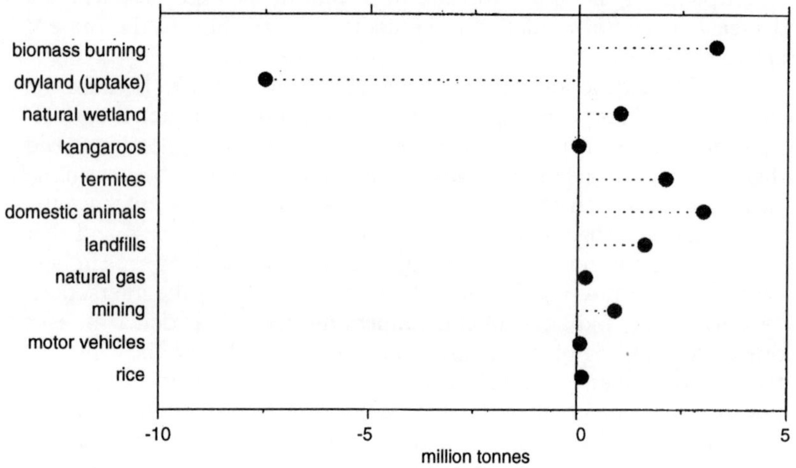

Figure 6.2 Annual methane emission and uptake by various sources in Australia
Castles, 1992

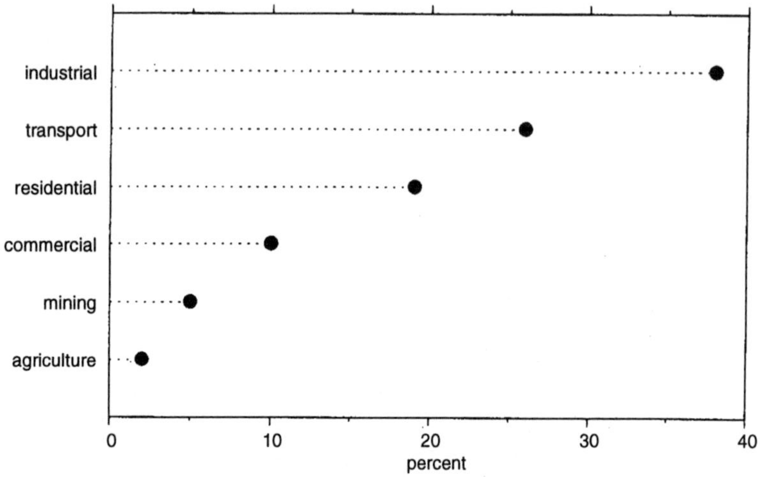

Figure 6.3 Carbon dioxide emissions from sectors of Australian industry
Castles, 1992

meltwater from the poles and causing unpredicted sea-level rises. Pittock (1989), one of Australia's leading greenhouse experts, expects more of the 'surprises' will be 'nearer to home' in the form of regional deviations from the average global trends, due to changes in the ocean currents around Australia, or to local fluctuations in cloudiness and rainfall caused by interaction between atmospheric circulation, local topography and distance from the sea.

With the scientists now preparing government, industry and the community for change (of an unspecified extent and period), planners have understandable difficulty in deciding on local pre-emptive strategies to meet the oncoming hazards. The same applies to the insurance and financial industries which base much of their decision-making on probabilities and forecasts. Statements such as the following from a leading expert place local authorities in an invidious position:

> We are likely to see, with greenhouse warming, whole sections of society in this country whose lives, property or livelihood will be threatened either through no fault of their own, or, as knowledge becomes available, through a failure to take the necessary precautions because of the cost or lack of planning (Pittock, 1989).

When Australia was admitted to the global program of the United Nations in June 1989, the citation read 'No other organisation or nation has so far conducted any comparable public awareness campaign on climatic change. Let's hope we're not "jumping the gun"'. In the meantime Australia is monitoring the sources of two of the important greenhouse gases to obtain better data on the origins and possible controls of these emissions, as shown in Figures 6.2 and 6.3.

Chapter 7

GRAZINGLANDS: 75% OF THE CONTINENT

This chapter introduces the potential and problems of the vast Australian pastoral zone. We examine how the inherently fragile arid zone became degraded as a result of the unplanned settlement of poorly understood ecosystems. Special attention is given to appropriate land tenure as a basis for sustainable production in these dynamic rangelands, so susceptible to drought, flood and fire. We examine the ways in which we can assess and monitor the changes in grazingland condition, as a basis for improving pastures and sustained cattle and sheep production. The chapter concludes with the search for the best of Australia's grasses for seed production and domestication.

7.1 Squatting — how to open up a new land

The grazing industry holds a very important place in the development pattern of land use in Australia. The way in which the squatters moved ahead of the land administration and local government led to the original location and boundaries of the vast sheep and cattle runs in New South Wales and Victoria. A somewhat more ordered allocation of land followed in South Australia and then in the 1860s in Queensland. One of the important aspects of the squatting era, which has been so well documented by Heathcote (1965) and Lines (1991), is how the view of land as an expendable resource has evolved in the Australian mind. The combination of the sheer vastness of the grazing country, endless plains and unpredictable climate seems to have developed in the landholder's psyche a view dominated by survival and the fight *against* nature.

While there is evidence of individuals who reflected an early sense of conservation and harmony with nature, the squattocracy and their employees were driven by a desire to take from nature whatever they could, seeing it as both a source of wealth and a threat to their very existence requiring constant vigilance. This relationship to the land led to over-use and ultimately to the degradation of the native pastures nationwide. Originally, the lack of drinking water in times of drought protected the great inland ecosystems of the Mulga, Spinifex, Saltbush and Mitchell grass country from overgrazing by native animals. In fact drought must be seen as the natural safety factor preventing the buildup of kangaroo numbers.

In the areas underlain by artesian basins, the successful drilling for underground water ushered in a new set of ecological controls in the inland pastoral country. Except for a few stations based on large virtually permanent waterholes such as occur along major rivers, it was the artesian bores and their great network of boredrains which made a permanent pastoral industry possible in the vast arid inland.

7.2 The instability and future of inland pastures

Elsewhere I have described the changes in the grass species and how this affects the pasture ecosystem (Roberts and Silcock, 1993). In essence, the more palatable grasses become over-used and replaced by hardier grasses and unpalatable weeds and shrubs. This change in what is termed 'botanical composition' reduces both the feed value and the density of the grass cover. The sparser ground cover then exposes bare patches of soil to wind and water erosion. The compaction of the soil surface has a twofold effect — it reduces the infiltration of rainwater and consequently increases the proportion of the rain that runs off. The outcome of this process is cumulative — drier soil and lower grass production — and at the same time it increases soil losses and landscape instability. Figure 7.1 shows the sequential steps in soil/water changes.

As the grass cover declines, the opportunity increases for woody weeds (encroaching shrubs) to spread and thicken up, especially in the absence of fire. Under natural conditions, the balance between grass and shrubs is maintained by a combination of competition from grass and the control of shrub seedlings by bush fires which are started by lightning and may occur every 3–5 years.

The widespread deterioration of the sheep and cattle pastures of the arid zone has led to several major studies of the processes and function of the native pasture ecosystem. In biological terms, the effect of overgrazing (too many animals) has been identified as the prime cause of degradation. Economically the inadequate size of grazing properties is regarded as the root cause of overstocking. The nationwide effects of clearing and grazing is shown in terms of 'level of disturbance' in Figure 7.2.

Some scientists believe that the hard-hoofed domestic stock are an unnatural and damaging element superimposed on an ecosystem that evolved under light grazing by soft-footed marsupials. Grigg (1988) was one of the first ecologists to evaluate the economic possibilities of a commercial kangaroo industry. With an area greater than half the size of New South Wales in danger of becoming permanent desert, the time

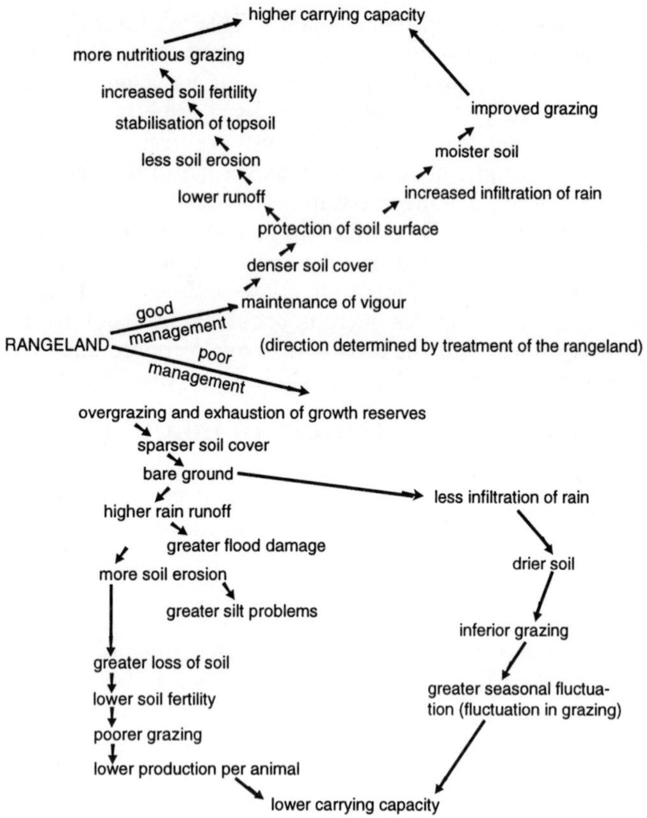

Figure 7.1 Relationships between management and carrying capacity of rangelands
Roberts, 1965

factor becomes vital to any large-scale rehabilitation strategy for inland pastures. The total removal of domestic stock would allow much degraded land to recover, but some alternative sources of income would be required if landholders are to continue occupying the arid zone.

Kangaroo meat is a healthy product and has become more available on the domestic market in recent years. Kangaroos are of course well adapted to the climate and feed of the inland region, but would have to be properly managed and harvested. This means balancing their numbers with the fodder potential of the drier regions. While the biological possibilities of a kangaroo industry are clear enough, the response of conservation groups to the idea of shooting Australia's national emblem is compounded by the overseas environmental lobby's opposition to killing attractive fauna. This is a good case study in selective emotionalism which overlooks the damage overpopulation of kangaroos is doing to the ecosystem. The concentration of high kangaroo numbers in the northwest sectors of New South Wales and Queensland is shown in Figure 7.3.

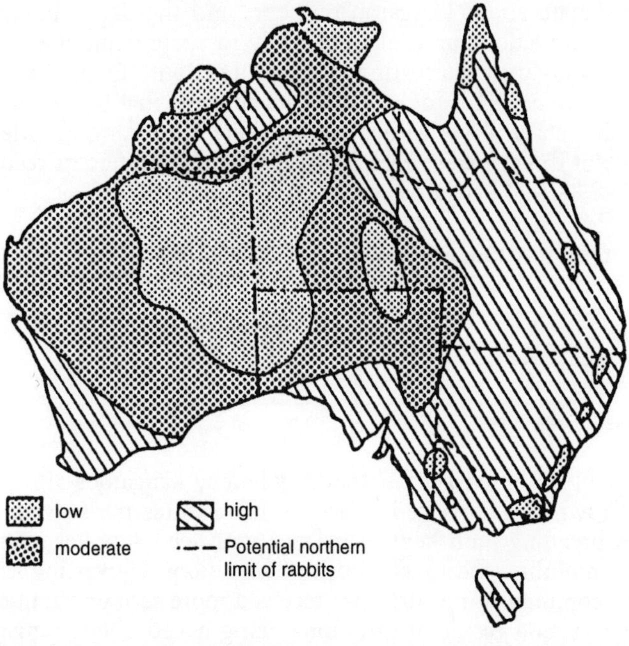

Figure 7.2 Degree of vegetation disturbance in Australia since European settlement in Australia
Anderson and Fox, 1982

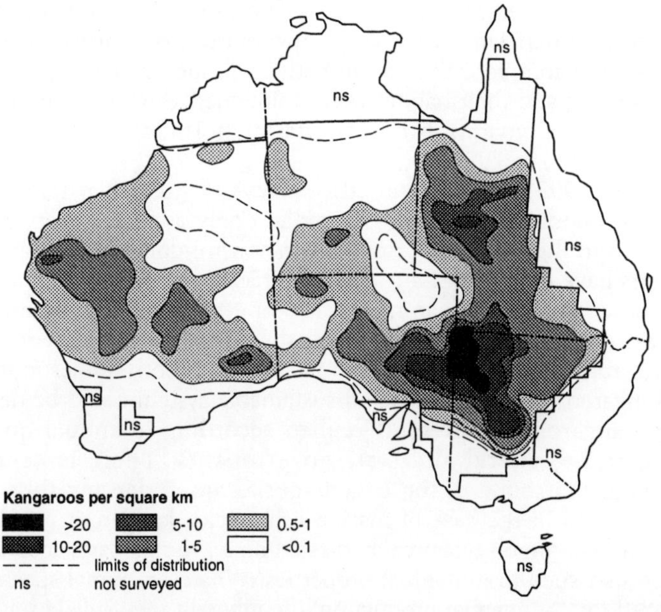

Figure 7.3 Estimated density and distribution of kangaroos in Australian rangeland
After Grigg, 1988, and Caughey et al 1987

The debate about kangaroo numbers and the dependability of kangaroo population surveys is not likely to abate while the acceptability of kangaroo harvesting is so contentious. Even if present estimates have a 50% error, present data suggests that the three main harvested species of kangaroos are nowhere near dangerous levels, certainly not threatened or rare. Table 7.1 gives the numbers recorded for the 1984–1990 kangaroo census.

Table 7.1 1984–1990 kangaroo census

Species	*1984*	*1987*	*1990*
Red kangaroos	5.7M	7.5M	10.4M
Western grey kangaroos	1.0M	1.7M	2.6M
Eastern grey kangaroos	2.9M	4.7M	4.9M

Source: Fletcher and Southwell, 1990

Grigg (1988) envisages a strategy whereby kangaroos are owned by the Crown but harvested in a way that ensures the survival of a vigorous breeding mob kept in balance with feed supplies irrespective of commodity (meat, skins) price fluctuations. Harvesting kangaroos as a commercial resource has received more serious attention in recent years as a result of questions being asked about rangeland degradation, costs of drought assistance to cattle and sheep producers, and increases in kangaroo populations. During the past 20 years the harvesting of kangaroos has met with national and international opposition from the animal welfare lobby. However, in 1990 the International Union for the Conservation of Nature (IUCN) endorsed the commercialisation of native wildlife harvesting, provided the species and their habitat were maintained. Similar support was given by the Convention on International Trade in Endangered Species (CITES) in 1993.

Grigg (1993) has made annual surveys of Australian rangeland kangaroo populations since 1978 and strongly advocates harvesting kangaroos in the wild, not farming them as semi-domesticated species as others have proposed. Kangaroo meat has been commercially available in the South Australian meat trade for several years and there are persuasive arguments that other states should follow. It is now recommended that if graziers can gain an income from both domestic stock and kangaroos, a more viable and sustainable system could be developed. Kangaroos are presently culled according to annual quotas agreed to by federal and state governments. There is general agreement that none of the culled species are under any threat of extinction. As in the case of marine fish stocks, kangaroos could be treated as a common resource, harvested on a pro rata basis according to permits issued to individual properties. Whatever animal species is preferred for commercial production, the inherent forage yield relative to rainfall will remain the prime determinant of carrying capacity, as shown in Figure 7.4.

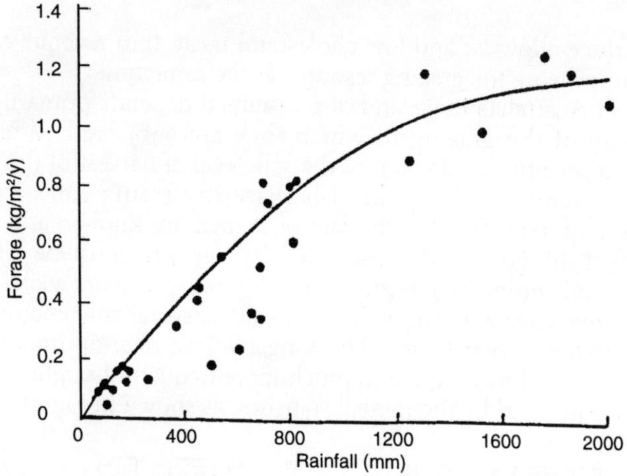

Figure 7.4 Relationship between annual rainfall and rangeland productivity
Wheeler and Freer, 1973

Sustainable rangeland management in Australia depends primarily on reducing the total stocking rate of all herbivores grazing the native pastures. There are financial incentives to reduce both sheep and kangaroo numbers, as reflected in the increase in the production per animal and improved drought-proofing of conservatively stocked properties. Whether kangaroos can be farmed in a system that involves mustering and yarding in specially fenced paddocks remains to be proven. Whatever method is used, it is necessary to overcome the cultural blockage of consuming Australia's national emblem and emphasise that eating kangaroo meat is no different from eating native fish, which has always been very acceptable culturally.

Presently kangaroos are regarded by graziers as a pest, not a resource. As such, apart from the monitored culling by professional shooters with permit tags, there is an unknown level of illegal and inhumane killing of 'the pest'. Proponents of commercial harvesting maintain that as a fully commercialised resource, kangaroo harvesting would reduce this inhumane activity to a negligible level. It is expected that when kangaroos are changed from being 'free loaders' into 'paying guests', landholders will have a much greater respect for them and their habitat as revenue-earners. Grigg (1993) estimates that the meat from more than 50% of the 3.5M kangaroos that are shot legally each year rots in the paddock, and believes that such waste is socially intolerable.

Sustainable rangeland management may involve a major change in the overall animal populations of the rangelands. Presently the arid and semi-arid regions carry about 25M sheep, 15M kangaroos, 6M cattle and 1M feral goats. The wool produced from these rangelands is about 14% of the national clip, while meat accounts for 22% of the national total (Freudenberger, 1993). The CSIRO estimated that in some regions each kangaroo was costing the grazier $20 in lost wool production. It makes sound economic sense to turn such animals, which produce a low fat and low cholesterol meat, into revenue earners while improving the grazing resource at the same time.

Whether Australia's rangelands are sustained depends primarily on the intensity of the grazing to which they are subjected. What is required is a scientific evaluation of the safe level of harvest of the natural fodder produced. The yield of herbage dry matter can be estimated, the requirements of each class of animal are known, and it is accepted that 40–60% of the available plant growth must be left to protect the soil. From long-term rainfall figures, the plant yield and thus the animal carrying capacity can be calculated reliably enough to form a basis for Crown leases. The longstanding over-optimism on stocking rates has, however, led to much more frequent drought declaration than is justified by the rainfall statistics, as shown in Figure 7.5.

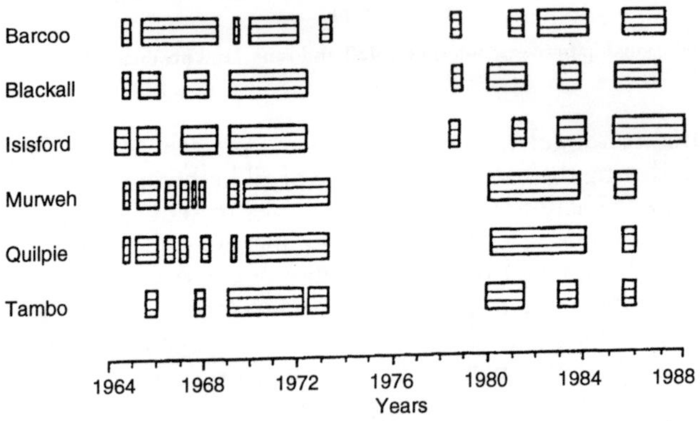

Figure 7.5 Drought declared periods for western Queensland shires (1964–1988)
Mills, Turner and Catabianco, 1992

Recent attention to practical methods of judging pasture condition and trend can greatly assist in achieving a sustainable basis for the grazing industry. With the advent of improved drought assistance policies, the management and maintenance of Australia's rangelands will, it is hoped, improve significantly. The future of Australia's inland pastures will largely be determined by government policy. There is a general belief that all land users should act as stewards of the land and most people consider that government land administrators should attempt to influence land use to benefit society. In the arid zone, governments try to influence land use by attaching covenants to leases. Theoretically any lessee who fails to comply with lease covenants should forfeit the lease. However, despite the threat of forfeiture, its application is rare, so the concept lacks credibility. As a result of recent inquiries into the matter, the Northern Territory, South Australia and Western Australia have all recommended introducing fines for lessees who fail to comply with lease covenants and conditions. As in the case of forfeiture, unless fines are imposed when covenants are breached, they will lose their credibility.

The control or regulation of land use is a sensitive matter and must be able to change with time. This means that the nature of covenants attached to leases should be periodically reviewed and, if necessary, changed. Any permanent form of leasehold tenure preventing such change must eventually lead to the development of more general controls and, if land users want a say in how these controls are framed, to some form of local government. If governments are serious in their attempts to prevent land degradation then the most efficient and appropriate method of achieving this is to introduce a land tenure and land management system that encourages pastoralists to conserve the lands they use.

Pastoralists cannot be expected to take an interest in conserving land for the future unless they are given a permanent interest in that future. This would involve granting them the security of a continuous form of tenure and yet being able to acquire other pastoral land as soon as a more appropriate use is identified for their present land. We must expect pastoralists to put their personal interests first and those of the land second.

The Australian arid zone is unique by global standards and places a special responsibility on administrators and managers to develop sustainable land use systems for its preservation. Compared to other global deserts, this region has been used for only a short time (about 130 years) and is the least populated by humans (less than 300,000 people on 5.5M km^2). Pastoralism occupies 66% of the arid zone and Aboriginal lands plus national parks occupy about 20%. The proportions of five states used for grazing and other purposes are shown in Figure 7.6. In the pastoral area, adjusting stock rates is still a major

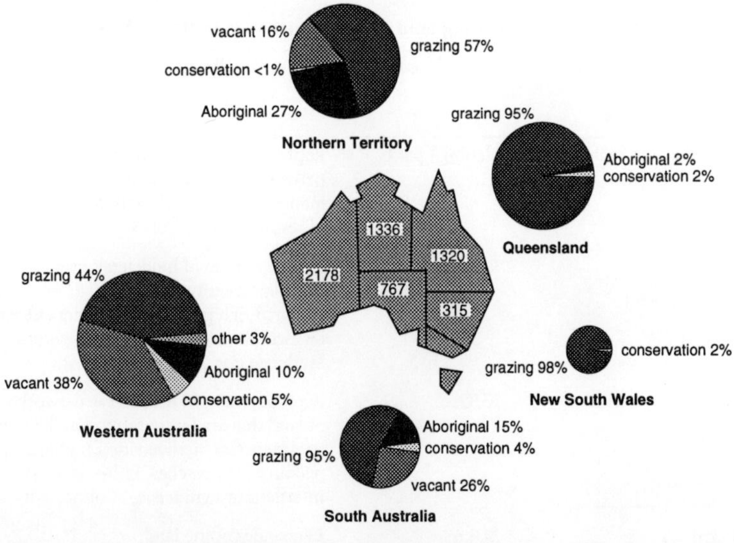

Figure 7.6 Percentages of land area in five states and territories used for grazing and other purposes, and the total rangeland area (in '000 km^2— central map)
After CSIRO, 1990

problem for landscape stability and continued productivity. Unlike Africa's lands, grazed by great herds of hardhoofed antelope and zebra (or North America's, grazed by bison), Australian grazinglands did not evolve under heavy grazing, but under soft-footed marsupials.

Today the ecologically sustainable management of the arid zone aims to maintain carrying capacity and biodiversity, and Morton et al. (1994) have suggested treating arid landscapes according to their major functional subtypes: resource-poor areas, resource-rich areas, and resource-poor areas that include fertile patches of soil and moisture. They propose superimposing four categories of land use on the mosaic of the three subtypes of landscape, as follows:

A National Parks (NPs)
B Excised Management Units (EMUs)
C Restricted Use Units (RUUs)
D Sustainable Use and Living Areas (SULAs).

These researchers propose the following principles for arid land allocation:

1 Identify areas with significant resource-rich patches and dedicate a proportion of them to conservation.
2 On the remaining country allow a range of other uses but manage all areas according to their needs.
3 Limit management elsewhere to whatever system is sustainable for the particular land use.

In their land use classification there are no absolute differences between NPs, EMUs and RUUs and the criteria for allocating land to each land use unit are shown in Table 7.2.

Table 7.2 Criteria for allocating land to four major uses in the arid zone

Primary spatial scales of land allocation and management	*Units of land*	*Criteria for allocation*
National and regional	National parks	Representation of biodiversity in reserve network. Large size and efficiency of management by park staff. Major aggregations of resource-rich patches.
Regional	EMUs	Representation of biodiversity through complementarity with parks. Emphasis upon resource-rich patches that cannot efficiently be incorporated into the park network. Moderate size. Connectivity.
Regional and local	RUUs	Representation in the reserve network of areas of land that are temporarily significant for persistence of biodiversity. Emphasis upon resource-rich patches. Either small size or intermittent significance. Connectivity.
Local	SULAs	Remainder of the land.

Source: Morton et al., 1994

7.3 Pastoral tenure and new criteria for management

Land tenure is only one way in which government can affect land use with a combination of rewards, penalties and incentives. Freeholding has come to be regarded as a type of tenure under which the state loses all control over how the land is used, while leasehold is seen as a tenure that constrains the landholder's freedom in land use decisions. Both these perceptions require qualification. In the past 16 years there have been land tenure inquiries: in Western Australia (1979), Northern Territory (1980), South Australia (1981) and New South Wales (1983) and in 1990 Queensland reviewed its leasehold legislation. It is of the utmost importance to the widespread adoption of pastoral zone landcare that land tenure is revised to reflect both the state's custodial role in resource maintenance and the landholder's role as land steward.

Land rights must be balanced by land responsibilities just as the 'right to farm' must be balanced by the 'responsibility to maintain the land'. Land use legislation in the past has given scant attention to protecting the land, at least as far as implementing the law is concerned. The complex nature of the socio-economic responses to land use legislation almost invariably results in unplanned outcomes to tenure reviews and sometimes the effects are the opposite of those intended.

One of the dilemmas of land tenure administration is how to define the elusive minimum viable size of property known as a 'Living Area'. Earlier governments in most states used this unit as the basis for selecting and balloting blocks in the initial stages of community development. It is clear now that small property size was a major cause of overuse and pasture degradation, and major problems arise whenever government policy continues to obstruct aggregation and opposes individuals operating several living areas. Both these matters are vital to the development of sustainable pastoral production. The prime goal of the state in determining appropriate land use policy should be to achieve a custodial role over Crown resources. Such achievement is most likely when property sizes and aggregations favour good management, not necessarily by a larger total number of landholders.

The previous Pastoral Act (1974) of South Australia required lessees not to overstock the land to an extent that would reduce the land's productive ability. In practice this led to a 10% increase in the average stocking rate for that state's pastoral zone over a 15-year period. It seems more practical for governments to base covenants on land condition monitoring rather than on actual stock numbers.

There are considerable problems with the clarity of land use legislation objectives in some states. Closer settlement was a universal element of all states' Land Acts, but according to Young (1985) only Queensland and New South Wales now pursue this objective through limits to the size of aggregations. Since about 50% of Queensland's sheep properties are below the size required for viable economic production, the Crown's policy as a major cause of land degradation is open to challenge.

Good land tenure systems stand or fall by the carefully considered balance between penalties (sticks) and rewards (carrots). There have been large discrepancies between what land laws make provision for, and how they are carried out. In democracies dominated by rural voting power, the imposition of penalties for non-conformity to tenure conditions has understandably been treated with great restraint. Similarly, the provision of low rentals and generous drought assistance have featured in such political situations. What is required is an agreement between the Crown and the lessee which gives financial rewards for improving the land (perhaps a reduction in rates) and penalties for degrading the land (perhaps a serious consideration of shorter lease renewal periods). Either way, some method of determining land condition and recording the results for later comparison is required.

Rural sociologists distinguish between tenure systems with economic implications and those which affect social status and self-esteem. If financial pressure is a major cause of overstocking, fines or other monetary penalties are unlikely to overcome degradation problems. Whether rewards in the form of reduced rentals could significantly encourage conservative management should now be considered. Subsidies in drought times may work well if they are based on previous conservative stocking.

As for social status and self-esteem, the introduction of rent-free leases may hold several advantages. First, collecting low rentals often costs more than it raises in funds for the state. Secondly, the psychological reaction of lessees 'getting their money's worth' is superseded by a more genuine basis for a stewardship agreement between the state and the landholder — that is, between the custodian and the steward. Young (1985) points out that the essential difference between term leases and perpetual leases is that the latter give special value to the grazing potential of the land. When a term lease expires, compensation is paid only for improvements constructed on the property — the condition of the land itself is of no consequence. At least in theory, a perpetual lessee can be compensated for having increased the carrying capacity of the land. Converting 30-year leases to perpetual leases can significantly encourage landcare if the covenants on the leases are appropriate. In addition, perpetual leases do not reduce the Crown's ability to keep future land use options open — an often quoted disadvantage of freeholding from the administrator's point of view. Perpetual leases were not issued in Queensland until 1968 and efforts to encourage good management by attaching covenants to these leases have been negligible. Such covenants provide the best opportunity to influence management in line with regional planning strategies and should be reviewed periodically (about every 15 years) as the vegetation and soil situation change. An advantage of such reviews is that leases could be reviewed for a shorter period if lessees had not met land use standards, and the original longer lease could be reconstituted once defaulters had corrected their practices. This variable security of tenure could provide an important incentive for future lease arrangements.

The goals of pastoral land administration have changed in the past two decades, but the traditional goals are still entrenched and reflected in most of the land legislation and administration. The major changes in these goals are summed up by Holmes (1990) as follows:

- Replacement of the pastoral development goal with that of sustainable use. Development goals were enforced by covenants on minimum stocking rates and minimum levels of capital investment. Sustainable use requires new covenants focusing on the maintenance of range condition and preservation of valued habitats. As with development goals, these are appropriately pursued through covenants attached to individual leases, with mechanisms for inspection and compliance.
- Replacement of the goal of conversion to more intensive agricultural use with the goal of accommodating other uses in association with continued pastoral occupancy. Goals of land use conversion were linked to those of closer settlement. Short-term leases enabled governments to pursue programs of land subdivision and conversion to higher-rent, longer-term tenures in an anticipated parallel progression of land use intensification and land tenure change. These assumptions of land intensification and closer settlement can no longer be sustained. However, the leasehold system can now be reformed to accommodate a variety of emerging land use demands of which the two most important are conservation/preservation and recreation/tourism.

If these changes are to come about within a reasonable time, both the administrative and biological aspects of the pastoral lands situation will have to be considered (Holmes, 1990):

- A very high proportion of Australia's rural lands are committed to pastoral pursuits, far exceeding their share of rural output. Pastoral leases comprise over 76% of all land held for private use, yet provide only an estimated 5% of the value of primary production.
- A similar relationship holds for individual pastoral holdings, which generally are measured in hundreds or thousands of square kilometres, yielding only modest returns relative to the area used.
- A high proportion of pastoral lands are fragile and susceptible to degradation.
- The risk of degradation is aggravated by extreme climatic variability, enforcing extended periods of excessive grazing pressure especially in droughts.

- Extremely low land values make corrective measures (other than reduced grazing pressure) and rehabilitation measures prohibitively costly. In low-value pastoral lands it is more urgent to emphasise preventative over restorative measures.
- There are very significant external factors involved in pastoral land degradation. The negative outcomes, present and future, arising from desertification; loss of vegetation, habitat and wildlife; catchment change; sand and duststorms; loss of amenity and loss of productive capacity, all suggest that the public should have a greater role in the management of our pastoral lands.

The conditions of pastoral land tenure will have to reflect more closely these changes in objectives and conditions, which in turn mirror the new community values of sustainable land use. Young of the CSIRO has given special attention to land tenure in the pastoral zone over the past 20 years, and has set out nine criteria which he believes are necessary for environmentally responsible land administration. Until recently, no state met more than two of these criteria, although South Australia has now moved closer to Young's ideal. The criteria are as follows:

- The covenants and conditions attached to leases must be periodically reviewed.
- All lease covenants and conditions must be simultaneously reviewed on a regional basis.
- The covenants attached to leases within each region must be reviewed about every 10 to 15 years in a manner that does not threaten pastoral security.
- Full compensation must be paid to the lessee whenever the government wishes to either (a) reallocate land to another use or person or (b) decrease the permitted intensity of use.
- Covenants must objectively state what is required of a lessee.
- Administrative actions must be taken whenever objectively collected data indicates that a substantial covenant breach has occurred.
- Appeals against administrative action taken in response to covenant breaches must make it possible for a court of law to determine whether, beyond reasonable doubt, a breach has occurred.
- Compliance with lease conditions must guarantee continuous pastoral use rights to all lessees, their heirs and their successors.
- By statute, non-compliance with lease conditions must remove the guarantee of continuous use rights until the breach is rectified. (Young, 1990)

The challenge in the 1990s is for policy-makers and administrators of Crown leases to tackle four neglected aspects of pastoral land use: (a) protecting biodiversity with appropriate lease conditions, (b) monitoring the condition and degree of degradation of leasehold land, (c) ensuring the proper management of Aboriginal land, and (d) recovering the costs of land administration from the users of that land.

7.4 Monitoring rangelands by condition assessment

The assessment of the 'state of health' of Australia's native pastures is regarded as one of the most important aspects of monitoring the progress of sustainable grazing management. Because of the enormous area to be monitored, the careful selection of techniques is essential. A good starting-point in evaluating how Australia should go about this vast assessment task is to consider who uses a monitoring system and what purposes each user might have. The Western Australian Rangeland Section sums this up as follows:

User	*Purpose*
The established pastoralist	As an assessment of his resource
The prospective pastoralist	As a regional overview
The extension worker	To prepare management plans
The resource manager	To establish monitoring programs
The research organisation	To locate specific interest areas
The soil conservation committee	To identify rehabilitation projects
The government	To improve land use policies
The land administrator	To administer land use policies

It can be seen from this range of users and purposes that several different objectives can probably only be achieved by several monitoring systems of varying intensity and scientific content. For instance, the established pastoralists may require regular seasonal inventories of the feed available in their paddocks as a basis for adjusting stock numbers and spelling those paddocks most in need of recuperation. Such monitoring would rely on a simple system that records the dominant fodder plants and the density of the plant cover by using visual estimates. The research organisation, on the other hand, would require a scientific technique for measuring, classifying and storing data on botanical and soil conditions to record trends in resource productivity. It would also require ready access to such data.

Various practical methods have been used to measure and record the vegetation in Australia's rangelands and several states have devised successful systems. It has been somewhat more difficult to produce an acceptable method for recording changes in the soil surface conditions of the native pastures. Table 7.3 presents the most recent proposal, by the Queensland Department of Primary Industries, of a method for non-scientists.

Table 7.3 Indicators of rangeland soil condition

Surface condition	*Indicators*
1 Stable	Soil surface stable and undisturbed
2 Slight disturbance	Soil slightly disturbed by erosive forces. Indicators include: • soil surface compaction or crusting • signs of increased surface runoff • signs of increased sheeting by wind or water • cattle tracks affecting water movement
3 Moderate disturbance	Soil moderately affected by erosive forces. Indicators include: • definite signs of sheeting or rilling • definite signs of soil mounding around obstructions • salt-affected areas • gravel and stone pavements common
4 Severe disturbance	Soil severely affected by erosive forces. Indicators include: • runoff concentrated in discrete channels • prominent sheet, rill or gully erosion • prominent sand mounds • vegetated areas isolated in mounds or depressions • scalds, exposed sub-soil • gravel and stone pavements prominent
5 Very severe disturbance	Soil very severely affected by erosive forces. Indicators are as for severely disturbed above, but more extreme.

Source: Pressland and Anderson, 1992

7.5 Sustainable stocking rates

The debate on how to achieve lasting productivity and stability for Australia's rangelands almost invariably reverts to the carrying capacity and stocking rate of each type of native pasture land. This is not to say that other factors such as the type of animal (cattle, sheep, kangaroos), the size of paddocks, distribution of water or thinning of trees are not important, but the level of harvest as controlled by the number of animals is the prime determinant of ecological stability.

It has been known for years that in many different types of pastoral country the reduction of animal numbers often leads to higher production per animal and even to higher total production per hectare. Economic comparisons also indicate that profits may be increased by carrying less stock. Because these animals are better fed, they grow better and can increase profit margins by lower costs and better quality of products. The basis for the carrying capacity of any pasture is the climate/soil environment in which it grows. In general terms, the higher the rainfall the greater the production of herbage. In the case of subtropical Australian rangeland types, Mott and his colleagues have summed up this relationship in Table 7.4.

Table 7.4 Relationship between rainfall and production of herbage dry matter (DM) in subtropical Australia

Vegetation type	*Dominant grass*	*Annual rainfall (mm)*	*Annual DM production (g/m²)*
Monsoon tallgrass	*Themeda triandra*	750–1100	112–270
Tropical tallgrass	*Heteropogon contortus*	600–1460	320–420
Subtropical tallgrass	*Heteropogon contortus*	700–1150	180–350
Midgrass on clay soils	*Dichanthium* spp.	500–900	200–300
Tussock grassland	*Astrebla* spp.	380–1800	100–230
Acacia	*Enneapogon* spp. and *Thyridolepis* spp.	180–500	70–120

Source: Mott, Williams, Andrew and Gillison, 1985

These data reflect the actual productivity and thus carrying capacity ranges of Australia's northern rangelands. It is within these productivity ranges that the effect of stocking rate (animal numbers) influences the stability of the grazed ecosystem, first through runoff, then through soil loss. Reduced cover usually increases runoff, and this in turn increases soil losses in most cases. The precise relationship between vegetation cover, runoff and soil loss depends primarily on rainfall intensity, slope and soil type.

The effects of increased runoff on the growth of native pastures are significant because when less rain becomes available to plant roots the amount of herbage and the species of plants change in time. The change in soil moisture means that xerophytic (drought-tolerant) species become dominant. These are usually less productive and less digestible than the original pasture species. As the runoff removes more topsoil the soil's fertility and growth potential decline. The relationship between erosion and pasture yield is shown in Figure 7.7.

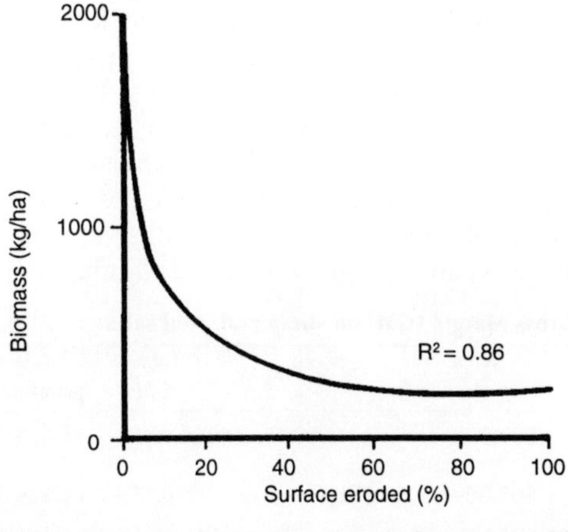

Figure 7.7 General relationship between erosion and pasture yield in Queensland Mulgalands

Mills, Turner and Catabianco, 1992

In this dynamic ecological situation the pastoralists must gauge the appropriate stocking level to gain the best production per unit area and to maintain a high rate of reproduction. While the effect of stock numbers on total wool production and lambing rate varies with the condition of the pasture, an example from arid Western Queensland over 25 years (Table 7.5) illustrates the importance of matching stock numbers to available feed.

Table 7.5 Stock performance on Offham Station, Cunnamulla

Period	*Stocking rate (Sheep/ha)*	*Wool production (kg/ha)*	*Lambing (%)*
1957–58	0.60	1.88	65
1958–59	0.65	2.77	61
1959–60	0.66	2.10	76
1979–80	0.47	2.40	91
1980–81	0.44	2.20	82
1981–82	0.42	2.35	93

Note: Two periods with similar annual (295 mm) and winter rainfall show the increase in wool production per hectare and reproduction rates with lighter stocking rates.
Source: CSIRO, 1990

The importance of this trend toward improving output by reducing animal numbers involves more than a short-term productivity benefit — it enhances the sustainability of the biological system as a whole. Overgrazing is prevented and soil protection is improved, runoff is reduced, organic matter in the soil surface is increased and better pasture species are encouraged.

Of equal importance to the biological stability of the system is the economic outcome of adjusting stocking rates downwards, especially in the drier regions of the inland. Wynne (1992) of the New South Wales Department of Conservation and Land Management has made a special study of the economics of stocking rates in the rosewood-belah country in the Ivanhoe-Balranald area. He found that pastoralists who stocked conservatively benefited in at least five ways: they had considerably higher wool cuts per sheep; at least the same total wool yield as more heavily stocked properties; higher lambing percentages; high stock sales; and lower stock losses. Table 7.6 is based on a survey of 20 pastoralists in the Western Division; the comparison of gross margins (= price minus direct variable costs) was obtained for three stocking rates of 4,000, 5,000 and 6,000 ewes run on 26,315 ha of native pasture with an average rainfall of 275 mm.

Table 7.6 Total Gross Margin (GM) on sheep and wool sales

Stocking rate	*Wool*	*Sheep*	*Total GM*	*GM per ewe*	*GM per ha*
4,000 ewes (conservative)	$67,900	$37,000	$104,900	$26.22	$4.01
5,000 ewes (typical)	$65,669	$31,400	$96,669	$19.33	$3.67
6,000 ewes (heavy)	$63,833	$26,000	$84,833	$14.97	$3.41

Source: Wynne, 1992

These data have particular significance for a sustainable wool industry in Australia. In essence, the conservative pastoralist was rewarded with higher gross margins per ewe and per hectare as a result of higher wool per sheep, better lambing rates and higher prices for sheep sold. In this case it was noted that as the stocking rate increased above the carrying capacity figure used for rental determinations (1 sheep per 3.5 ha), the gross margin per sheep and the profit per hectare declined. It was deduced from this study that conservative stocking can double the profit earned by landholders with moderate debt levels. Wynne (1992) points out that heavy stocking is leading not only to lower wool yields per sheep and poorer lambing rates, but also to 'ecological losses' reflected in a reduced capacity of properties to withstand drought.

The ecological requirements of native pastures are reflected in the effects of various intensities of grazing, that is, various levels of use or what is termed 'percentage utilisation'. In practice, the 'take half - leave half' rule of thumb has been recommended as an initial guide to help grazing managers decide when to move their animals from a paddock. This guideline suggests that the paddock should not be grazed bare but that at least 50% of the herbage should remain for conservation. The Queensland Department of Primary Industries at Charleville has studied the effects of the intensity of grazing on native pastures in the Mulgalands of Western Queensland. Their results for four levels of grazing are shown in Table 7.7. Similar results were obtained in Mitchell grass pastures in the same region and it was concluded that irrespective of rainfall, lighter grazing results in better cover and better long-term yield.

Table 7.7 Effect of grazing intensity on grass density

Utilisation	*Basal cover % (density of grass at soil surface)*		
%	*1978*	*1979*	*1981*
20	1.7	2.1	1.1
35	2.0	2.0	0.7
50	2.6	2.1	0.3
80	2.8	1.9	0.3

Source: Orr, 1981

Stocking rates must be considered in the broadest possible context of sustainable land use. The future of Australia's rangelands, especially the arid zone, depends on several interrelated factors as diverse as stocking-rate appraisal, kangaroo harvesting and tourism promotion. Stafford-Smith (1994) has noted that the value of income from alternative land uses for the whole rangeland region for the financial year 1991–92 was:

- cattle $527M (17% of the Australian total)
- sheep and wool $296M (4% sheep, 12% wool)
- tourism $3B (8%)
- mining $10B (38%).

Traditionally the outback districts have been cattle and sheep country and the state administration has responded to the need for land use policy in a variety of forms of land tenure, stocking rate controls and drought assistance arrangements, as shown in Table 7.8.

Table 7.8 Differences between states in tenure, stocking and drought policies

State	*Tenure*	*Stock constraints*	*Drought declaration*	*Drought assistance*
New South Wales	Mostly perpetual leases	Notional	Recommendation of Pasture Protection Boards (pastoralists)	Transport subsidies
Northern Territory	Perpetual leases, no review	None	Property by property, by government inspection	Capped freight and loan subsidies, 2nd declared year only
Queensland	Very varied: being rewritten	Notional	Property by property, but shire declared if more than half its properties; regional drought committee with pastoralists	Freight subsidies for fodder, water carting and return of stock, and sundry waivers/ concessions
South Australia	42-year lease with 14-year review	Set maximum (10-year average)	None	None
Western Australia	Leases, to expire 2015, being replaced	Notional minimum	Property by property, based on rainfall analysis only	None: declaration is psychological mainly

Source: Stafford-Smith, 1994

The striking differences between states all administering the same type of resource are remarkable: 'The state by state tenure laws have involved an extraordinary duplication of legislative effort and angst only to achieve inconsistent outcomes. It would seem time to rationalise and unify such policies ... to minimise the legislatively imposed barriers to sustainability' (Stafford-Smith, 1994).

The debate on land use in the pastoral zone and the stocking rates that form a basis for sustainable animal production is essentially to do with accepting the concept of stewardship (Roberts, 1993). For land stewardship to be fair and lasting, there must be agreement on the private rights of the individual (personal benefit) and on the responsibility toward the community (common good). Where virtually all the benefit of productive land accrues to the landholder, it is reasonable that he/she bear the costs of maintaining the land resource. This would be the case in good cattle country not ruined or visited by substantial numbers of tourists. In other cases, where the land is less productive and very little profit accrues to the individual landholder, but the landscape draws large numbers of tourists, it would be fair to expect that industry pay the costs of maintaining the area.

7.6 Domestication and the search for adapted grasses

Australian researchers and advisers tend to prefer exotic grasses and legumes for improving pasture quality or revegetating denuded areas. In the US about 45% of the 82 common grasses used for reseeding rangelands are native to that country. Up until very recently, Australia had no registered native cultivars available in the seed trade, despite having 750 native grasses to select from. These grasses have adapted to the harsh climate and infertile soils of this continent and are thus suited to the low input production systems preferred for sustainable production. To be useful for domestication, economical to harvest and easy to establish, the species would have to meet a range of criteria. Lodge and Groves (1990) suggest that in addition to their use as animal feed, there is a growing market for adapted grasses for a range of non-agricultural uses:

- at roadsides, for stabilisation; in favour of introduced species which tend to invade adjacent native bush
- in local government parks and recreation areas because they incur low maintenance costs
- by soil conservation authorities, for soil stabilisation
- by river control authorities to stabilise creekbanks and dam foreshores
- by mining companies, for large-scale minesite rehabilitation
- for gardens, amenities and landscaped areas.

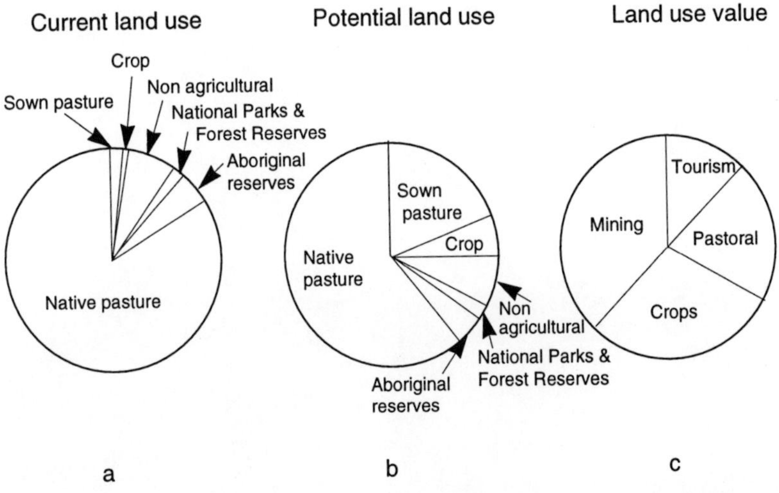

Figure 7.8 Potential for intensification of land use in Australian savanna lands

Tothill, Nix, Stanton and Russell, 1985

Australia must put more research effort into exploiting the untapped potential of its vast genetic pool of grasses. An expansion of the type of work done on *Danthonia linkii* (Lodge, 1987) would be a very good investment for land stabilisation. Using the open woodlands (savanna) region of Australia as an example, Figure 7.8 shows the extent of the potential to expand the area under sown pastures.

While the shortage of moisture has always been recognised as the ultimate limitation for rangeland production, various strategies to increase dry matter yield have been attempted. These include ripping, ponding, pitting and seeding those grasses that can make the best use of accumulated surface moisture. Where acute deficiencies of one or more plant nutrients limit herbage yields, fertiliser has been added but the cost, particularly of transport, has usually made this form of intensification uneconomic.

Future research is needed to apply the work on *Danthonia* to a wider range of the better native grasses. In some cases all that is required in the initial stage is a simple but large seed collection project as is being done with Mitchell grass (*Astrebla*) in northern New South Wales and bluegrass (*Dicanthium*) in Queensland. The author (Roberts, 1993) has proposed a list of 41 of Australia's 111 indigenous grass genera that are worth further evaluation for seed production and pasture reinforcement.

Chapter 8

FORESTRY, TOURISM AND MINING: OPTIONS FOR PLANNERS

This chapter examines several alternative land uses apart from agriculture: forestry, tourism and mining. We consider good multi-purpose forestry as an improvement on the old single-purpose 'tree farming' in our high rainfall catchments. We analyse the ecological limits to tourism and explain how Australia can become a world leader in ecotourism if we plan ahead and protect the ecosystems on which this industry is based. Finally the chapter explains the situation in the rehabilitation of minesites, with special reference to opencut coal mining as a major land use in three states.

8.1 Forestry as a sustainable land use alternative

Any consideration of sustainable land use in Australia must include forestry as the most appropriate alternative. Forestry usually makes its strongest claim as the prime land use option in humid climatic regions where mountainous terrain makes sustainable agriculture impractical. This is not to suggest that gently sloping or open rolling country is not well suited to timber production, but cropping and grazing are usually more appropriate for flatter country.

The geographical distribution of Australian native forests and exotic plantations indicates that traditionally the land settlement policies of all state governments encouraged the clearing of all regions where topography suited broadacre agriculture. This large-scale removal of trees was extended to the semi-arid regions of the Victorian and West Australian Mallee, and the drier scrub country of central South Australia and New South Wales. In more recent times further development took place in central Queensland where 1.4M hectares of Brigalow (*Acacia harpophylla*) was cleared in the late 1960s.

Forested landscapes serve important purposes apart from timber production: water supply catchments, national parks and tourism. Many of Australia's most prized wilderness areas, including World Heritage areas such as the Wet Tropics of North Queensland, in fact generate much more income from tourism than would be possible to sustain by any other use. Australia's major timber forests consist primarily of the hardwoods (*Eucalyptus*) in Western Australia (jarrah and karri) and in southern New South Wales and southern coastal Queensland. Besides these major timber resources are the cypress pine open forests of northern inland New South Wales and central western Queensland. Although much of the timber industry in the rainforests of North Queensland has been discontinued in recent years, some tropical forestry specialists believe that logging could be carried out indefinitely in these humid forests if the correct strategies were applied.

Timber extraction techniques can usually be classed into three types:

- Clearfelling, a technique whereby the entire population of millable trees is felled and either replanted or naturally regenerated.
- Selective logging, where only trees of a prescribed size (usually measured as diameter at breast height) are felled, and care is taken to minimise the damage to surrounding trees during log extraction operations.
- Coup logging, in which designated squares or rectangles in the forest are clearfelled except for selected seed trees which are kept for regeneration.

Forestry research indicates that from any of these methods permanent well-managed timber resources can be gained, provided that soil losses due to water erosion are minimised by correct extraction procedures. Such procedures include the environmentally sound planning of forest roads (for log transport and fire-fighting) and the reduction of runoff by stacking brushwood on the contour (along the slope). The role of fire in the regeneration of some important species such as jarrah is also an important aspect of forest management in some regions. Many of Australia's hardwood forest areas have been cut-over several times since the early days of the timber industry. These regrowth forests continue to produce commercially viable volumes of millable timber, often on the basis of a 30–50 year rotation, depending on the diameter of logs required. In the northern rainforests the timber yields are relatively low (40–60 m^3/ha) but the loss of nutrients incurred by removing timber must still be accounted for in sustainable logging. In tropical forests, soils are generally acidic and low in nutrients, although it is often mistakenly assumed that their lush growth indicates high soil fertility. In fact, most of the nutrients (up to 80%) in such ecosystems are contained in the trees themselves, rather than in the soils, as occurs in grassland or in cropping systems. In theory, the rate of pedogenesis (soil formation) from the bedrock underlying the forest should make nutrients available at least at the same rate they are being removed by logging, if the system is to be sustainable — that is, unless the nutrients are supplemented with fertiliser, which is common practice in plantation forestry but not in timber-getting from native forests. Charley (1991) has studied nutrient cycles on the sandy forest soils of Fraser Island to gain a better understanding of how nutrients are lost by rainforest logging.

From a sustainability viewpoint, the soil types and their pedogenesis rates are all important when estimating how much timber can be removed and how much erosion can be tolerated. Rainforest, for instance, grows on a range of red, grey and yellow earths derived from both tertiary (recent) basalt volcanic rock and from ancient metamorphic (granite) rock. The younger basaltic soils are considerably more fertile than the granitic earths, while the latter are more erosive and prone to more severe soil loss when the tree cover is removed from steep slopes.

The Queensland Department of Forestry suggests that the northern rainforests can be maintained as a permanent timber resource by applying a harvest rotation of not less than 40 years, at least in many of the better stands in far North Queensland. There are, however, serious differences of opinion between tropical forest specialists on whether any form of commercial logging can be truly sustainable in terms of maintaining the original structure and function of the rainforest ecosystem. For this reason the Federal Department of Environment and Heritage has moved to discontinue all logging in the World Heritage (Wet Tropics) area and has compensated the timber industry accordingly.

8.2 Good silvicultural practice and multiple use of forest land

Logging plans specified for hardwoods differ from those for rainforests although basic similarities may be found in all environmentally oriented timber extraction operations. Because the tropical rainforest has received so much media attention (if not scientific attention) in recent times, the recommended procedures for Crown land timber extraction as described by Nicholson (1984) warrant explanation. The trees to be logged are marked by the departmental field officers, giving due consideration to minimising damage to remaining trees and leaving sufficient replacement trees ('growing stock'). A logging plan is prepared, showing roads, loading ramps, major snigging tracks (log haulage routes), creek crossings and buffer strips. This plan becomes legally binding once agreed to by the Crown and the purchaser of the logs. Trees to be harvested are marked with a numbered blaze to indicate the direction of fall. Selected trees marked for retention must be avoided by the loggers, who are fined if avoidable damage is done to them. The forms of environmental damage and the proportion of the forest area affected have been analysed by Keys (1984), as shown in Table 8.1.

Table 8.1 Type of disturbance from rainforest logging tracks and ramps as percentage of total area

Disturbance category	*%*
Vegetation removed	1.8
Topsoil exposed	1.3
Topsoil disturbed	4.0
Subsoil exposed	1.1
Subsoil disturbed	6.8
Total disturbance	15.0

Source: Keys, 1984

Our forests are now more commonly regarded as a multiple-use resource, not just for timber, tourism and catchment purposes but for a wide range of other human activities. The problems of balancing the requirements of each use with those of the others is fraught with difficulties. Allocating forests for complementary purposes such as nature reserves and wilderness values can usually be accomplished, but activities that involve noise, traffic or large numbers of people in a forest detract from its natural beauty and tranquillity.

8.3 Tourism and its ecological limits

Chapter 2 referred to tourism, forestry and mining as the main alternatives to agriculture in land use planning. The section on forestry explains how tourism can be combined with other activities in forest land. In Australia tourists tend to seek different types of experiences in a range of landscapes. The 'beach culture', while attractive to many visitors, has virtually no impact on agriculture as a competing land use. However, in many coastal areas where beautiful scenery and a mild climate attract tourists, the competition between the 'holiday industry' and agriculture can be intense. The north coast of New South Wales, the Margaret River area of Western Australia and the Sunshine Coast of Queensland are cases in point. Tourism has expanded significantly in the past 20 years to become one of Australia's three top earners of foreign exchange. Increasing numbers of overseas and domestic holiday-makers require more accommodation facilities and more transport infrastructure such as roads, airports and bus stations, all of which encroach on established rural production. A range of intensive production lines have been affected, e.g. fruit, vegetables, dairying and sugar cane.

Many tourists from cities prefer the rare scenic beauty and tranquillity of unspoilt wilderness areas, and this preference has led to a remarkable expansion of tourism based on ecological appreciation, now termed *ecotourism*. It seems that this growing appreciation of natural environments is universal, perhaps because people realise that our planet's untouched ecosystems are becoming scarcer.

Australia has a range of dramatically different natural environments for tourists to study and enjoy — from Ayers Rock in the harsh 'red centre' to the Daintree Rainforests in the humid tropics. Tourism in such places and others like the Tasmanian southwest and the Northern Territory's Kakadu has a very strong claim for being the most appropriate sustainable land use. This claim can be justified on economic and environmental grounds and as such has gained favour with planning authorities over grazing, forestry, cropping and even mining in some instances.

Some primary producers still maintain that tourism is not a genuinely productive industry like crops, grazing, timber or minerals, but this claim is losing ground as the contribution to the local economy from tourism not only approaches but in some cases exceeds the income from land-based production. Table 8.2 reflects the growth and extent of tourism in economic terms since 1985.

Table 8.2 Growth of tourism in Australia

Number of international visitors:	2.1 million (1989–90)
Average annual growth rate of international tourism:	14% per year since mid-1980s
Average annual growth rate of domestic tourism:	2% per year since mid-1980s
Contribution of international tourism to GDP:	\$6.7 billion (1990) — 9% of Australia's total export earnings
Number of people employed in tourism:	5.7% of workforce (1989–90)
Origins of main short-term international visitors, 1990 (in descending order):	Japan, New Zealand, other Asia, UK/Ireland, Europe, USA

Source: Bureau of Tourism Research, 1991

The effects of tourism on the land mostly depend on numbers of people, but the type of activities tourists pursue must also be taken into account. Thus tourism differs from other land uses in that while it doesn't remove materials or energy from a region, it imposes direct effects such as trampling, litter, sewage, dust from tour buses and noise from aircraft. If tourism is to be a permanent enterprise in Australian natural environments, it will need to be strictly regulated to keep it at a level that these ecosystems can tolerate. In practice this means tourist operators sharing the costs of infrastructure and amenity maintenance with the local authorities responsible for shire development.

Though not constrained to the coastal zone, the tourist industry is largely concentrated in the subcoastal belt, which puts extra pressure on this narrow intensively developed strip of land. As a case study in how tourism affects land use planning, the coastal zone includes the following environmental hazards:

- building construction on sand dunes
- illegal clearing of dune and riparian vegetation
- uncontrolled access to fragile ecosystems
- overfishing
- subdivision of wetlands for development
- destruction of heritage sites
- illegal dumping of wastes
- water pollution
- trampling of reefs and other fragile ecosystems.

To overcome this complex set of human-related problems, government (a) regulates allowable tourist activities, including limits on visitor numbers, (b) uses economic instruments to encourage minimal impacts, and (c) draws visitors' attention to damage risks and appeals to their sense of responsibility. In an attempt to combine these three 'management tools' into a workable program for the tourist industry, a national inquiry into tourism was made in 1993. This inquiry — a good example of how the environmental effects of an industry can be identified, evaluated and controlled — led to the National Tourism Strategy (CZI, 1993) which included four strategic goals: environmental, economic, social and industrial. The strategy aims to provide for sustainable development of the coastal zone in a manner which preserves ecological and heritage values while optimising national income and employment from tourism. It also aims to ensure favourable social outcomes from tourism for local people while enhancing tourism experiences for visitors. Finally the strategy provides for research, coordination and promotion in support of the future tourism industry as one of Australia's top foreign exchange earners (Department of Tourism, 1992).

The National Tourism Strategy is perhaps the best Australian example of how a successful planning and management strategy can be constructed as a basis for sustainable development. The other industries featuring in the national ESD investigation (agriculture, mining, forestry) have yet to produce a national industry strategy. The OECD (1991) has specified the characteristics of a successful strategic approach to planning sustainable land use, stressing that the inclusion of environmental, economic and social goals is essential. The National Coastal Strategy is based on an approach that builds on the OECD guidelines. It emphasises processes rather than an 'end state'; it allows for the evolution of programs as experience is gained; it takes community values seriously; it resolves competing objectives; it forms policy by consulting with all stakeholders; it encourages local participation in policy-making; it recognises the conditions of land capability as limits to development; it provides for enforcement of land use guidelines, and it includes monitoring and evaluation as a basis for modification and improvement.

It is worth noting (for application to other industries) that the National Tourism Industry Inquiry recognised four spheres in which *widespread reform* was required, namely:

- Deficient planning mechanisms (for approvals and coordination).
- Ineffective translation of policy into practicable management guidelines.
- Lack of a community-based management that employs local understanding and commitment to sustaining the industry. This is expressed as a need to 'empower' communities.
- Defective integration of policy and procedures by all government levels, departments and units.

For the balanced development of Australia's agricultural, mining, forestry and fishing resources, the current defects in planning mechanisms and the proposed reforms all warrant serious consideration.

8.4 Minerals and mine-site rehabilitation

The popular image of mining is one of great devastation that leaves huge craters and turns pristine landscapes into bare moonscapes. There are several good reasons why the environmentally unfriendly image of this industry should be changed:

- Minerals, Australia's top export earner, bring in about $20B annually.
- Mining operations affect less than 0.1% of the nation's land surface.
- Modern rehabilitation techniques have made minesites even more productive after mining than they were before mining.
- Several opencut mining companies have developed Australia's biggest and best revegetation programs.

Governments have recognised that both sequential land use (forestry, then mining, then grazing) and multiple land use (forestry, recreation and tourism simultaneously) can be achieved in many situations. In the case of mining, environmental impact assessments are required before operations commence. Also, in most states mining companies are required to deposit a rehabilitation bond which is returned only after the authorities are satisfied that the minesite rehabilitation has been adequate. The Mining Act or its equivalent in all states requires that mining companies restore the land, as closely as possible, to its original (pre-mining) condition. This original condition includes landscape contours (topography), topsoil and vegetation type and diversity. In practice the mining industry's rehabilitation record is far from perfect, but examples abound of excellent rehabilitation under conditions as different as the Cape York and Western Australian ecosystems. The most impressive programs are those of the bauxite miners (shallow strip mining of lateritic soils) at Weipa in Queensland and in the southwest of Western Australia. Other large-scale rehabilitation has taken place in the opencut coalfields of the Hunter Valley of New South Wales, the La Trobe Valley of Victoria and the Bowen Basin of Queensland. In all these areas the topsoil is first stockpiled, then returned after mining once the overburden (shale and sandstone) has been levelled to at least an 8% slope.

With modern rehabilitation methods, which may cost about $25,000 per hectare to apply, the mined land reverts to excellent grass/legume pastures or to productive commercial forestry. Mining companies' environmental sections have relied heavily on the agricultural researchers' recommendations for their topsoiling and replanting techniques. The methods advised by soil conservation authorities form an important part of the rehabilitation program. The hydromulching of batters (spraying a mixture of seed, clay, fertiliser and organic matter) has been widely applied, as has the use of contour banks and waterways. Improved grasses and legumes have given many minesites a considerably higher carrying capacity than the original grass could achieve.

Despite these positive outcomes of mine-site rehabilitation, there are still serious environmental problems caused by some mines. Not all companies are committed to being responsible corporate citizens, nor do all state governments administer their own Mining Acts diligently enough. There are serious local problems of heavy metal pollution, acid mine drainage, soil erosion and dust on some mines, and more stringent application of the conditions of the mining leases is required.

Chapter 9

ISSUES, PRINCIPLES AND PROGRAMS FOR AUSTRALIAN AGRICULTURE

As the heading implies, this chapter takes us through the fundamentals of our agricultural policies. We deal with the symptoms of unsustainable production systems and contrast these with the 'principles of permanence'. We examine present programs which aim to improve our landscapes and see how they encourage the achievement of our objectives of sustainability in crop and animal production. The chapter ends by briefly considering the Australian population policy as a key element in planning and resource exploitation.

9.1 The issues — symptoms of unsustainable systems

Ann Hamblin and her colleagues on the Working Group on Sustainable Agriculture (WGSA, 1991) give Australia credit for leading the world in developing sustainable practices such as pasture legume rotations. They have also catalogued 14 major degradation issues, listed their causes and suggested remedies (see Table 9.1), which they qualify by pointing out that these remedial ('ameliorating') practices do not cover all degradation issues, nor do they work in all situations. Although the issues here have been identified in collaboration with state land use authorities, this whole sphere requires more research to develop a rigorous methodology.

Table 9.1 Causes and control of sustainability issues in Australian agriculture

Issue	*Contributing practice*	*Ameliorating practice*
Decline in soil nutrients and biological activity	Rotations that lack grain/pasture legumes.	Improved rotations with wider range of alternative crops to control soil pathogens and maintain fertility.
	Inadequate soil/plant testing for nutrient levels.	Adequate testing and choice of appropriate fertilisers/rotations.
	Insufficient/inadequate fertiliser use.	Management of pastures for maintenance of soil fertility and biological activity.
Soil structural decline	Excessive cultivation.	Minimum tillage.
	Bare soil and fallowing practices.	Stubble retention.
	Overgrazing and loss of groundcover.	Use of integrated rotation and grazing management for cover and weed control.
	Animal/machinery traffic on wet soils.	Matching tillage and stocking to soil type and condtion. Use of gypsum on degraded soils.
Soil acidification	Non-use of lime on acid soils.	Regular lime applications.
	Use of acidifying fertilisers.	Use of appropriate fertilisers.
	Use of shallow-rooting pastures.	Use of deep-rooted perennial pastures where possible.
Soil erosion	Poor cultivation techniques. Degradation.	Use of deep ripping, minimum tillage, pasture rotations to restore fragile soils. Use of appropriate earthworks to control water flow.
	Overgrazing. Insufficient vegetation cover.	Retention of cover by stocking adjustment, wildlife management, stubble and roughness retention. Use of windbreaks to control wind erosion.
	Poor matching of enterprise to capability of land.	Improved land capability assessment.
Water quality	Inadequate waste/effluent disposal systems in intensive enterprises.	Improved engineering for effluent disposal and animal housing. Provision of adequate health inspection procedures.
	Contamination of surface and groundwaters by fertilisers and pesticides.	Care in pesticide usage and application methods near open waters.
		Optimise chemical usage to reduce accessions to groundwater.
		Change in fertiliser type and application to improve uptake by plants.
		Application of fertiliser according to soil analysis.
	Sediment and salt runoff into surface waters.	Management to minimise soil erosion and secondary salinity.
Soil salinity Waterlogging (irrigated)	Inefficient/excessive water use • by flooding • too frequent irrigation	Improved water scheduling. Conjunctive re-use of groundwater.

Issue	*Contributing practice*	*Ameliorating practice*
Continued		
Soil salinity Waterlogging (irrigated)	• through low infiltration.	Drainage and gypsum to improve infiltration.
	Inadequate deteriorating infrastructures.	Improved water distribution networks (infrastructure).
	Poor site selection for irrigation areas.	Site selection consistent with soil and land capability.
Soil salinity Waterlogging (dryland)	Excessive clearing of deep-rooted perennial native species causing rise in groundwater levels.	Identification and revegetation of recharge areas. Strategic tree and shrub planting/ management. Use of deep-rooted perennials wherever possible. Farm-plan implementation to preserve trees and shrubs.
Pesticide residues and resistance	Over-reliance and persistent use/ overuse of pesticides. Over-reliance on chemical control of crop weeds.	Whole-farm integrated pest management. Biological control of pests. Selection of genetically resistant plants and animals. Low pesticide use farming. Biodegradable pesticides. Development and use of vaccines. Use of rotations to reduce pest/ weed/pathogen burdens.
Vegetation degradation	Over-grazing. Poor use of grazing management to control weeds in pastures.	Stocking rates consistent with land capability. Improved grazing/stock management to gain maximum pasture benefit. Improved management for drought. Utilisation of soil fertility as it builds up.
	Poor weed control.	Adequate weed controls and quarantine measures.
Remnant vegetation decline	Insufficient fencing to protect remnants. Stock pressures on young/ vulnerable stands.	Refencing on land-use basis to exclude stock. Reduced grazing pressure. Strategic tree/shrub planting.
Fire management	Insufficient and excessive use of fire in certain grazinglands.	Appropriate use of fire to maintain native grasses and to control woody weeds and shrub regrowth.
Feral and native animals	Inadequate control of feral and native pest animals.	Control of feral pests.
Consequences of crop monocultures	Reliance on a single crop without rotation.	Use of 'break' crops and pastures in rotations.
Land use competition	Lack of consultation in planning and appropriate dispute resolution. Inadequate or inappropriate land use planning.	Proper consultation in planning. More appropriate dispute resolution.

Note: This table represents an attempt to summarise on a technical basis some of the practices that can be detrimental to or enhance sustainability. It is not intended to suggest that those practices listed as 'good' will be appropriate in all circumstances, nor is it intended to set out policy prescriptions. It does not cover all issues nor all the complexities of the issues.
Source: Working Group on Sustainable Agriculture, 1991

9.2 The principles of permanent agriculture

For setting new directions for Australian agriculture, it is useful to develop a set of objectives or *principles* by which future trends may be gauged as appropriate or otherwise for sustainable agriculture. The following principles have been proposed for gauging production systems (WGSA, 1991):

First Principle: Farm productivity is enhanced over the long term.
Second Principle: Adverse impacts on the natural resource base of agriculture and associated ecosystems are ameliorated, minimised or avoided.
Third Principle: Residues from chemicals used in agriculture are minimised.
Fourth Principle: The net social benefit derived from agriculture is maximised.
Fifth Principle: Farming systems are flexible enough to manage risks incurred by the vagaries of climate and market.

In this book we shall refer to this set of statements as *Hamblin's Principles* and shall use them as a reference point for evaluating systems. The principles are best used within a framework of the *limits to disturbance* that form the boundaries of sustainability, beyond which agricultural production systems may lose their resilience and could collapse. Hamblin (WGSA, 1991) has illustrated the differences between systems which can absorb shock (disturbance) and those which cannot (see Figure 9.1). The shock may be ecological (e.g. toxins or soil loss); or economic (e.g. profit margins or equity levels). Systems with a greater capacity to recover can absorb a larger degree of shock without collapsing.

Moving from the Australian situation to the global situation as it relates to permanent agricultural systems, several requirements of permanent systems can be identified. First, permanency is encouraged by a more thorough incorporation of natural processes such as nitrogen fixation, nutrient cycling and pest-predator relationships in the farm production processes. Second, permanent systems will use fewer off-farm inputs with the potential to harm the soil, environment and health of the farm and its operators. Third, permanency will come from making greater and ecologically sound use of the biological and genetic potential of plants and animals in the spheres of disease resistance, conversion efficiency, moisture use and adaptation to weed competition. Fourth, permanent systems will closely match cropping patterns to the inherent limitations (susceptibility to degradation) of the landscape used for production. This means taking land capability data seriously when deciding on the intensity of land use. Finally, permanency will be improved when future farming policy places the emphasis on profitability and efficiency only in the context of preserving the productive potential and stability of soil, water and biological resources, through ecologically based management.

Permanent agriculture is not a single 'best' system but it includes

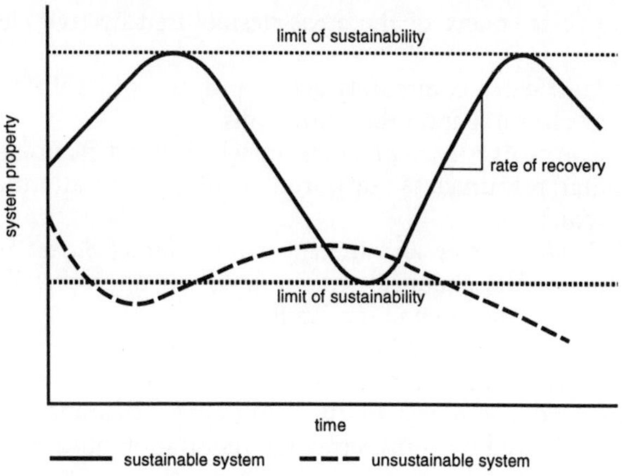

Figure 9.1 Diagrammatic comparison of production systems of different sustainability
After Working Group on Sustainable Agriculture, 1991

a range of combinations of basic principles and the practices based on them — for example, crop rotations that reduce weed, pest and plant disease problems. If legume nitrogen can replace fertiliser nitrogen and if agronomic practices and tillage methods can reduce soil loss to acceptable levels, the basis of permanency can be created. Added to these 'pillars of permanence', integrated pest management that uses scouting (bug-checking), weather forecasting, resistant crops, timing of planting and biological control can greatly increase the sustainability of cropping.

9.3 Current land improvement programs

While much work remains to be done to ensure permanence in Australia's rural production, the Commonwealth and the states have made laudable progress at least in establishing policies and programs as a foundation for environmentally appropriate land use into the next century. Some of the more significant federal programs are as follows (in no particular order):

- *National Conservation Strategy*, which aimed to offer a basis for preserving the nation's living resources. This was followed by State Conservation Strategies, at least in most states.
- *National Soil Conservation Program*, which aimed to involve landholders and authorities at all levels in improved land management and the adoption of a land ethic.
- *Murray-Darling Basin Management Strategy*, which coordinates catchment and water management in Australia's most important irrigation area, in a program designed primarily to control wetland salinity problems.
- *One Billion Trees Program,* which concentrated on revegetating on

a massive scale, many of the areas destabilised by tree clearing nationwide.

- *Greening Australia*, a community action program of local revegetation projects in rural and urban situations.
- *Water Resources Assistance Program*, in which good planning and use of water resources is supported, with special attention to salinity prediction.
- *National Residue Survey* and the *Australian Market Basket Survey*, which monitor the chemical residue levels in unprocessed food crops and in packaged retail foodstuffs.
- *Agricultural and Veterinary Chemical Council*, responsible for evaluating the toxicity, persistence and environmental hazard of new and registered chemicals in rural industry. Clearance by the Council is followed by considering the registration of each chemical by state authorities.
- *National Health and Medical Research Council*, which advises all user organisations on the health risks of chemicals.

Within the states there is a very wide range of resource protection programs and agricultural improvement projects which combine to form a network to guide locally adapted land management. Each state has a Landcare Plan for the Decade of Landcare (1990–2000), and programs such as the following are also being set up in various states:

- Trees on Farms
- Salt Action
- Whole Farm Incentives
- Timber Industry Strategy
- Agroforestry
- Native Vegetation Retention
- Responsible Use of Agricultural Chemicals
- Land Resource Assessment
- Integrated Catchment Management
- Applied Rangeland Ecology
- Extension Strategy
- Farm Financial Counselling
- Integrated Pest Management
- Rangeland Management
- Remnant Vegetation Protection Scheme
- Farming for Agriculturally Sustainable Systems
- Farmcare
- Community Nature Conservation

These programs are having varied success in different parts of Australia. The attempt to improve management practices by promoting one or more programs in a particular region has revealed that much more than innovative technology is required to attain widespread acceptance of new techniques. It is also apparent that there are no guidelines for planners confronted with the central issue in regional planning — the size of the target population of people for whom the plan is intended. Let us consider population as a factor in the planning formula.

9.4 Program planning for how many people?

Consider first the trend in the nation's population growth rate. The current Australian growth rate of 1.2% per annum is one of the highest in the western world. Population growth is affected by fertility rate (family size), death rate (infant mortality and longevity) and by the number of immigrants accepted annually. Immigration has been at a rate of about 120,000 annually over the past decade. Since 1945 about 25,000 Australians have emigrated each year, so the nation is gaining about 100,000 new people annually. The Australian population could be stabilised (zero population growth) at about 20M by the year 2075 if immigration equalled emigration and if family size was kept to 2.1 children (Cocks, 1992). The contribution of immigration to the Australian population for the four decades after 1946 is shown in Figure 9.2.

For an analysis of the Australian population trends, natural increase must be separated from immigration gains. Since 1947 the average annual rate of population growth has ranged from about 2.5% to 1.2%. Table 9.2, summarising the 1947–1986 period, reflects the proportion of increase accounted for by migration.

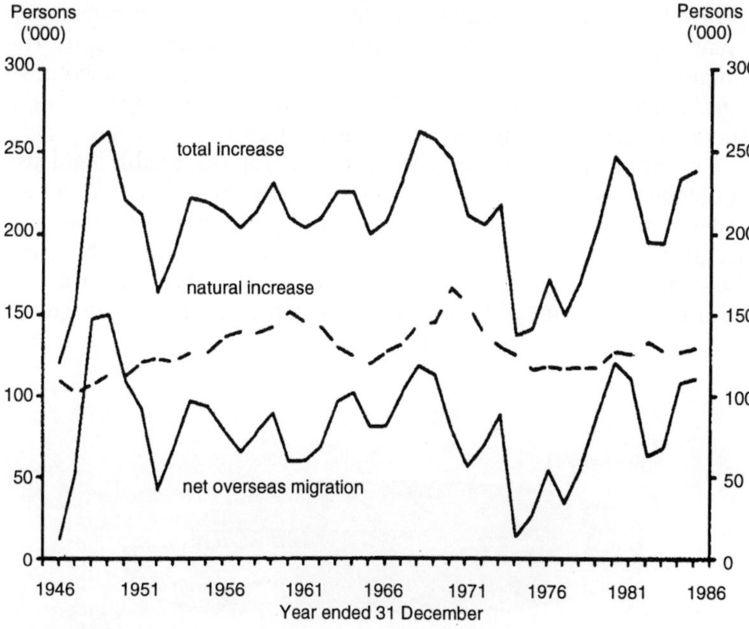

Figure 9.2 Components of Australian population growth (1947-1986)
Castles, 1992

Table 9.2 Population growth, 1947–1986

Intercensal periods	*Average annual growth ('000)*	*Total (%)*	*Average annual rate of growth* *Natural increase (%)*	*Net overseas migration*
1947–54	201.0	2.58	1.42	1.16
1954–61	217.4	2.35	1.44	0.91
1961–66	208.5	1.98	1.24	0.74
1966–71	231.2	2.01	1.16	0.85
1971–76	193.2	1.44	1.02	0.42
1976–81	178.0	1.33	0.82	0.51
1981–86	219.0	1.40	0.84	0.56

After Australian Bureau of Statistics, 1986

The Report of Fitzgerald (1988) suggests that even trebling the present immigration rate would add only about 10% to GDP per capita by 2030. Fitzgerald does not analyse the environmental effects on the landscape, which is unfortunate since regional planning must be firmly based on the land's *carrying capacity*, which in turn depends on rate of production, consumption and pollution. Gifford (1975) suggests that by doubling Australia's irrigation area, the nation could feed 60M people. In practice, the biological productivity of a region doesn't limit its human carrying capacity because commodities are imported from other regions. But landscape, space and the protection of nature do affect the quality of life, and the growth of population density increases stress and environmental damage. Despite the humanitarian arguments to increase immigration and to encourage large families, the overall effect of population growth jeopardises the Australian lifestyle, environment and economics.

The implications of population growth for sustainable land use are obvious. Unless planners can assume a target population for a particular region, weighing the alternative development plans against sustainability criteria is not possible. For instance, in the early 1970s when the federal government proposed growth centres such as the Bathurst/Orange and Albury/Wodonga complexes, there were guide-

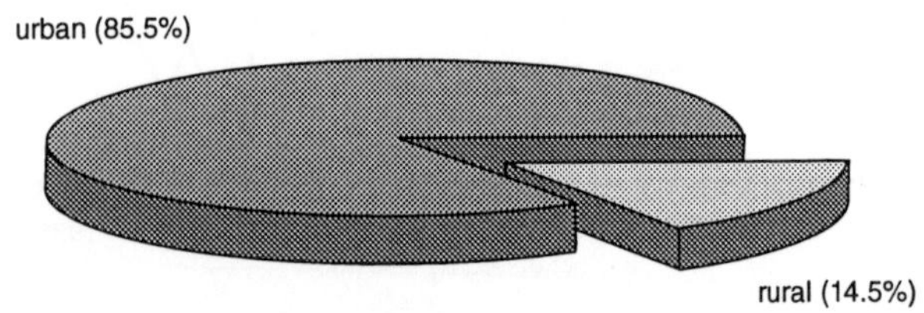

Figure 9.3 Urban and rural population of Australia (1989–1990)
Dept. of the Arts, Sport, the Environment and Territories, 1991

lines for infrastructure planning, services, food and water supply, plus energy demand forecasting. Several locations have been suggested for new cities (e.g. Queensland's central coast), but the basis for planning remains vague and uncertain. This precludes applying the concept of carrying capacity to human population density. The present concentration of the Australian population is shown in Figure 9.3.

So what assumptions should we make in planning our sustainable societies, if they are to be based on sustainable land use and fed by sustainable agriculture? In practice we separate the two elements of land capability planning from regional economic development planning. In other words we distinguish sound production practices for agriculture, pastoralism and forestry from other elements of development, and plan the intensity of rural land use according to the capability of each portion of land. Provided the physical and biological requirements for sustainable production are met, other elements of regional development can be planned on the basis of importing into the region the necessary production inputs for economic growth.

Chapter 10

PERMANENT PRODUCTION AND INTEGRATED AGRICULTURAL SYSTEMS

Chapter 9 dealt with the principles of permanent production, and this chapter turns to the technical aspects of integrated farming. We look at the factors that work against the adoption of sustainable agriculture techniques. We consider the need for a more long-term approach by our researchers, then we explain the full meaning of 'integrated production systems'. We conclude by scrutinising organic farming and the importance of soil organisms and their role in permanent cropping.

10.1 Permanent agriculture — what's the recipe?

Since the 1930s many writers have promoted good production systems, using terms like 'conservation farming', 'balanced farming', 'organic farming' and more recently 'holistic resource management'. First came the concern over soil loss through erosion, later the issue was fertility decline and the loss of organic matter. Today the focus is on chemical toxins and soil physical conditions, along with the earlier issues. The International Federation of Agricultural Producers (1991) drew up a useful list of the components of a modern sustainable production system:

1 The system should be *non-disruptive* of sociocultural networks and should avoid practices that infringe on accepted community norms.
2 The system should encourage *self-reliance* by using local knowledge and local resources as far as possible.
3 The system should be *appropriate* to local needs, abilities and finances and should be adapted to the local environmental conditions.
4 The system should be *resilient* enough to cope with shocks and extremes of weather, economics and biological threats.

5 The system should be *productive* enough to meet family needs and supply excess produce for sale.
6 The *profitability* of the system should generate sufficient return on capital and labour and provide sound continuous investment.
7 The balance between inputs and outputs should make the system *regenerative*, so that its mineral and nutrient levels are permanently maintained.
8 By not over-exploiting natural resources, the system should be *stable* by maintaining soil condition and biodiversity, thereby keeping options open for future generations.

Observers in developing countries have indicated the need for a clear definition of sustainability that deals with improving agricultural productivity so that it can keep pace with the demand created by population growth. The level of productivity to be sustained must be specified. Provided this level is within the capability of the land resources concerned, it should maximise productivity. However, if a country is already harvesting its renewable resources at or close to their full potential, it is unrealistic to extend the goals of sustainable agriculture to a point beyond their capacity simply because population growth demands it. Mother Nature has strict limits.

10.2 External constraints to achieving sustainable agriculture

The reasons why farmers are slow to change to sustainable systems are many and varied, but in general they are of two types: first, economic reasons which prevent producers breaking out of a production cycle supported by financial incentives from government; and second, psychological reasons which concern fear of risk, fear of innovations, and insecurity about change. As an example of the many factors requiring adaptation when farmers change systems, Figure 10.1 shows the nine elements of conservation tillage.

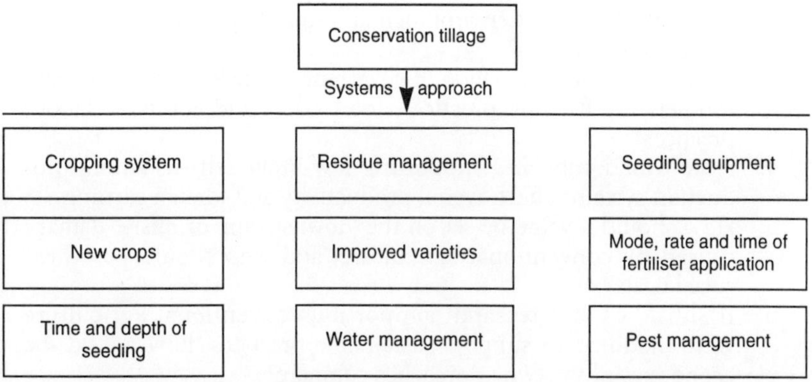

Figure 10.1 Cropping factors requiring consideration when adopting conservation tillage
Edwards, 1990

At a landmark conference at the University of California (1990) entitled 'Sustainable Agriculture: Balancing Social, Environmental and Economic Concerns', the external factors influencing the adoption of permanent agriculture were well catalogued. The most useful outcomes of this analysis were not the answers on sustainability problems, but the questions it asked on the external factors constraining biologically sound farming practices at the time. Eight issues were identified:

- *Economic policies* affecting commodity support, credit allocation, trade and taxes.
- *Concentration* of agricultural production in fewer but larger agribusiness owners.
- *Government subsidies* making producers depend too much on predetermined government choices.
- *Externalised costs* of resource depletion, ground water contamination and farm workers' health care imposed by chemically-based agriculture.
- *Comparisons* of 'sustainable and conventional' methods of production, which give no reliable indications of real cost differences.
- *Equal access* to pesticide-free produce not yet available, thus disadvantaging producers who use alternative methods.
- *International trade* and Third World debt leading to rural poverty as a result of forced export agriculture in subsistence societies.
- *Free market* economic pressures from short-term financial goals, discouraging sustainability.

Among the questions raised by this analysis, certain fundamental issues persist concerning serious constraints on the widespread establishment of sustainable agriculture:

1. How do agricultural policies affect incentives for producers wishing to apply crop rotations as fertility and pest management strategies?
2. How does the concentration of farming in the hands of large agribusiness firms affect environmentally sound practices and socially just labour practices?
3. How would the elimination of government subsidies improve environmentally friendly practices, food safety and equitable labour practices?
4. How would subsidies supporting low-input self-sustaining production systems affect overall productivity and foreign earnings?
5. How should a value be set on the 'downstream' or offsite damage caused by conventional agriculture, and who should pay these social costs?
6. If subsidies and research supporting conventional agriculture were modified to support sustainable practices, how would the economics of the two approaches compare?
7. How could changes to international trade policies encourage sustainable family farming on a global scale?

8 How could changes in food and social policy increase demand for, and access to, safe food while encouraging sustainable production?
9 How can family farms be encouraged to apply sustainable methods by changing free market operations which presently constrain competition?
10 Is the free market the best economic system to promote sustainability, or is the present marketplace inappropriate for protecting the long-term productivity of the land?

10.3 The need for a new research perspective

The US National Research Council's (1989) report, *Alternative Agriculture*, highlighted the environmental and human costs of high-input agriculture. The report argued that research should focus on alternative practices and modifications in technology and 'know-how'. It also called for research on the social, economic and institutional policies that influence farmers' choices of beneficial alternatives. Agricultural research is a subset of the broader natural resource management policy which in turn includes the research themes shown in Figure 10.2.

The past decade has indicated a growing recognition by agricultural and environmental research groups that they share common goals and should not be competing for mutually exclusive goals. Sustainable agriculture is now viewed as an integral component of sustainable

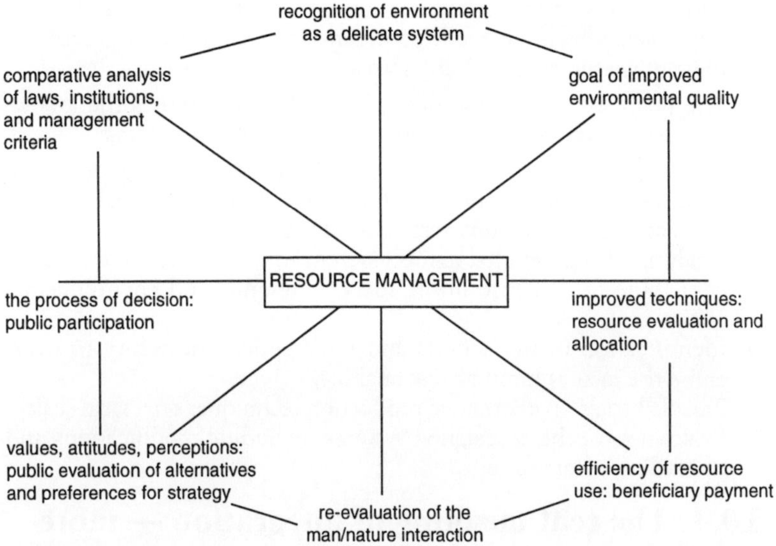

Figure 10.2 Interrelated research spheres in land resource management planning
O'Riordan, 1971

development, so research into appropriate future crop and animal production systems should incorporate the following characteristics:

- long-term maintenance and enhancement of biological productivity
- minimal environmental impacts
- adequate economic returns for farmers
- optimal production with minimised chemical inputs
- satisfaction of human needs for food and income
- provision of rural communities' social needs.

All definitions used by research planners thus include environmental, economic and social goals. They also imply *agroecosystem* flexibility, to cope with changes and stress; these attributes are well explained by Edwards (1989) with his emphasis on *integration*. Indeed, the lack of *systems research* has been identified by the National Research Council as the key obstacle to adopting alternative farming practices, and as the major step necessary to develop sustainable agriculture in the US (NRC, 1989).

Research programs for maintaining agricultural and other natural resource potential require interdisciplinary projects that include at least four elements:

1 Understanding the *components* and *processes* that determine the *functions* and *interactions* of the agroecosystem as an ecological whole.
2 Developing *hypotheses* that encompass all the system's interacting components.
3 *Testing* the validity of these hypotheses in the practical functioning context of the system.
4 *Interpreting* the experimental results in terms of each component and in terms of the agroecosystem, to overcome the lack of *interrelatedness* which has made so much previous agricultural research meaningless in sustainability terms.

The Board of Agriculture of the NRC (1991) proposes the following conceptual framework for research on integrated systems:

1 Description of the target agroecosystem, including its goals, boundaries, components and functions, plus its internal interactions and its interactions across its boundaries.
2 Analysis of the detailed agroecosystem to determine which factors contribute to or constrain its social, economic and environmental goals.
3 Identification of the actions and interventions necessary to overcome the factors limiting sustainability.
4 Practical trials on alternative production techniques on a farm scale.
5 Evaluation of the redesigned systems, including modifications and upgrading where necessary.

10.4 The real meaning of integration — more than a nice idea

The past half-century has seen more and more intensive cropping develop into today's high energy/high fertiliser/high pesticide crop production programs. The high yields of grain from these systems depend on continued high inputs into crop varieties bred primarily for high yield. Unless there is a change in the view that the elimination of small family farms and the degradation of the environment are acceptable prices to be paid for exponentially increasing grain yield, the ultimate biological collapse of our monocultures can be virtually guaranteed. A comparison of 'natural' and high-input systems is given in Figure 10.3.

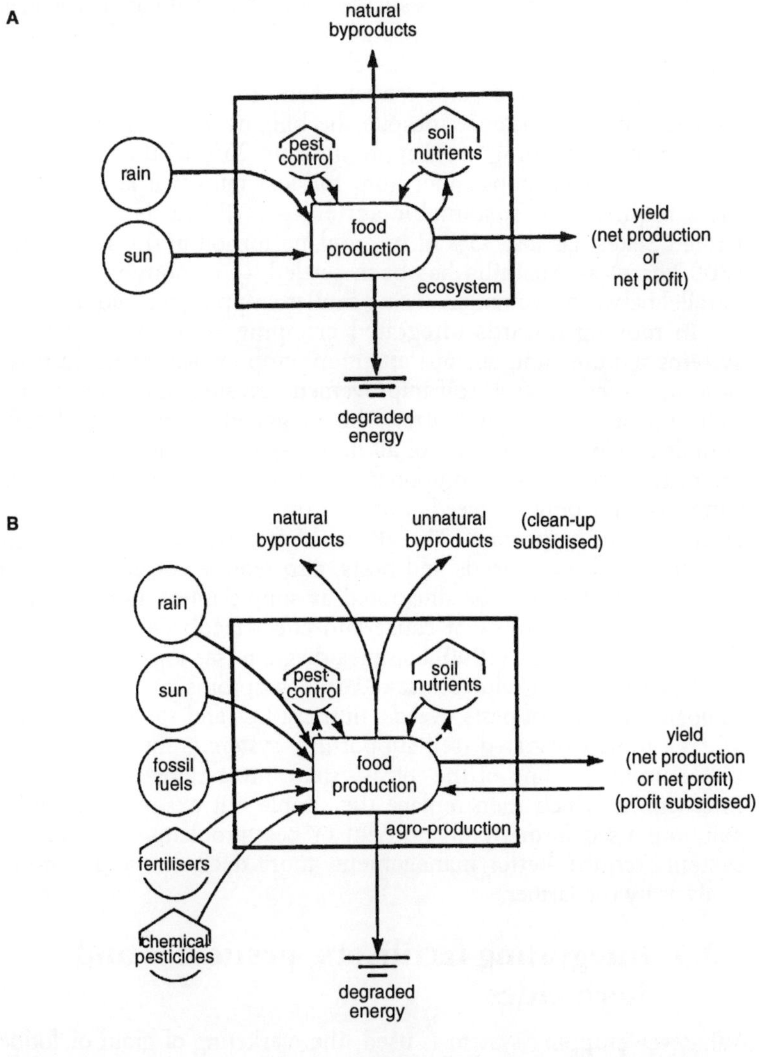

Figure 10.3 Diagrammatic representation of (A) a natural (unsubsidised) agricultural system and (B) a high-input (subsidised) system

Edwards, 1990

Overproduction and price collapse are two of the more obvious outcomes of this drive for ever-higher yields. The energy crisis of the mid-1970s also reiterated the disproportionate increase in energy consumption associated with these higher yields. Papendick (1987) has estimated that in the United States today, for every calorie of food crops produced three calories of energy are used to produce it, and seven calories are used to process and distribute that single food calorie.

Apart from the high chemical and energy consumption of these systems, the side-effects of surface and groundwater pollution, plus the destruction of wildlife (pesticides) and their habitat (herbicides), have accumulated to the extent that the whole 'high yield mindset' is being reconsidered. Besides this chemical degradation there is the physical loss, first of organic matter, then of the topsoil itself. In the United States for instance, despite the billions of dollars invested in soil conservation programs, approximately 25% of the 421M acres (170M ha) under crops suffer from soil loss rates that far exceed the rates required for sustainable agriculture (Edwards, 1990). The United States has lost 33% of its cropland topsoil in the same period (200 years) as Australia has been settled — a thought-provoking parallel between two nations using similar cropping techniques.

In moving towards integrated cropping systems, all the subsystems require simultaneous attention: crop choice, plant nutrients, pest and weed control, soil improvement, erosion reduction, market indicators and pollution control. The integrated system is much more complex than a simple sum of all its subsystems — it consists of the interactions between its components. For instance, the effects of cultivation — not only on weeds but also on plant diseases and organic matter — are important. The effects of fertiliser, not only on crop growth but also on weeds and pests, also require consideration. The integrated system can be illustrated as simple interactions or, more correctly, as a complex of cause-and-effect feedback responses, as depicted in Edwards' (1990) complex diagram (see Figure 10.4).

Chemical inputs have masked the symptoms of breakdown in biological control of pests, weeds, nutrient loss and soil erosion. Thus when a more integrated (self-supporting) system is initiated, a much better understanding of the intra-system interactions is required — interactions which then replace the simple but expensive chemical solutions used formerly on nutrient or pest problems. So integrated systems require better management, more decisions and complex analysis by the farmer.

10.5 Integrating fertilisers, pesticides and herbicides

Whatever cropping system is used, the marketing of grain or fodder crops represents an export of plant nutrients from the farm. If this level of nutrient export exceeds the rate at which new minerals are being made available by soil processes, a net loss of nutrients will

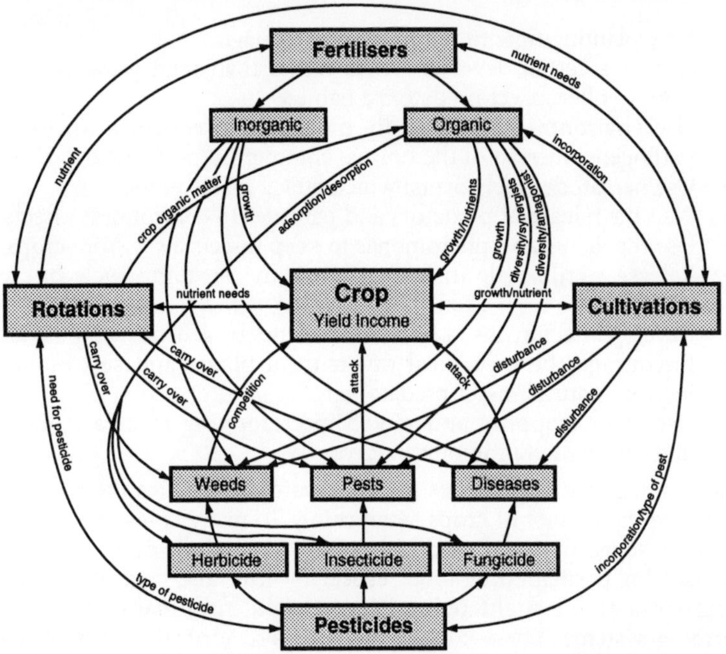

Figure 10.4 Complex interrelations between the components of modern cropping systems.
After Edwards, 1990

result if they are not added, either as mineral fertiliser, organic fertiliser, manure or legume nitrogen. Other sources may include sewage sludge, organic industrial wastes or other byproducts. Another approach to meeting the needs of crops is to breed varieties adapted to lower nutrient levels, and to develop with genetic engineering non-legume grain crops which can fix atmospheric nitrogen biologically.

The prime difference between so-called 'organic farming' and conventional farming lies in the use of agricultural chemicals. Of these chemicals the use of pesticides is the most important single component of the agroecosystem which alternative agriculture aims to reduce or even eliminate. This means replacing the synthetic biocides with other pest and disease control measures. The integrated system has a range of techniques for this purpose. Some of those proposed by Edwards (1990) are as follows:

- Reduce insecticide use by employing the best pest population forecasting methods.
- Apply pest control formulations strategically, so that lower dosages can be used.
- Apply crop rotations to avoid the seasonal carryover and buildup of pests of a particular crop.
- Select crop varieties (cultivars) that are less susceptible to pest attack.

- Vary planting time to minimise pest damage.
- Accept a certain level of weeds rather than total weed suppression, to offer insect predators a habitat.
- Use environmentally friendly biological insecticides (including pathogens) instead of the normal commercial pesticides.
- Use nematodes (eelworms) which attack insect pests.
- Plan the release of predators and parasites to control pest insects.
- Use repellents and pheromones to keep insects away from crops.
- Release sterile male insects to break the breeding cycle of crop pests.
- Develop insect toxins in new crop varieties by genetic manipulation.
- Encourage the biological diversity of plants and soil biota to increase natural insect predators.
- Use strip cropping and mixed intercropping to create habitat diversity, which promotes biodiversity.

These integrated approaches to pest control can reduce the costs and improve the quality of crops significantly. In many cases one or more of these suggestions are used in combination and the premium usually paid for pesticide-free food, especially fruit and vegetables, often compensates for slight reductions in yield from many 'biological' farming systems. These control methods also form the basis of integrated pest management (IPM) which is now becoming the accepted approach to pest control. For new system designs, pesticide flowpaths in the ecosystem must be considered, as shown in Figure 10.5.

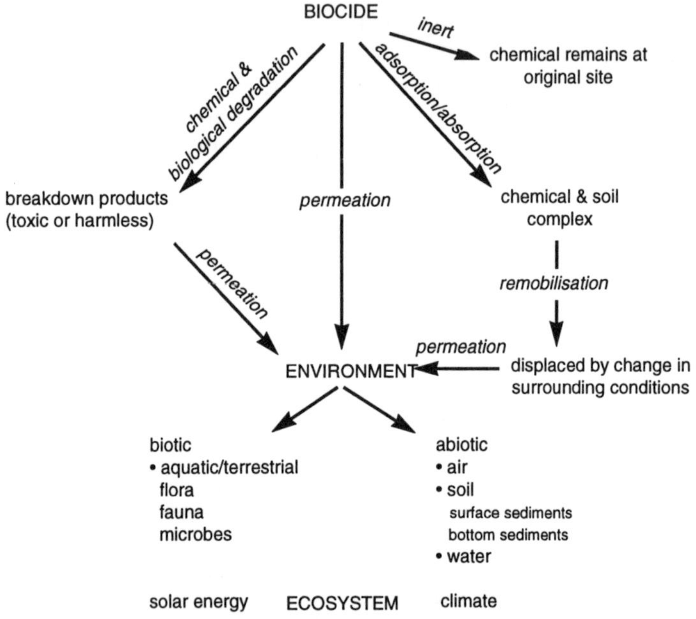

Figure 10.5 Pathways of pesticides and herbicides through ecosystems
Crone, 1990

10.6 Soil biota — organisms at the interface of integration

Most soil organisms contribute to the vital processes of decomposing and mineralising (making available) organic matter in the soil. As such, they regulate soil fertility but also play an important role in building and maintaining soil structure through their exudates which bind soil aggregations, and through their physical mixing and aeration of the soil.

Because of their important but largely unrecognised contribution to sustainable agriculture, the groups and actions of soil biota warrant explanation. Basically these organisms can be divided into plants and animals (flora and fauna), with each of the size categories contributing to 'soil health' in different but complementary ways, as shown in Table 10.1.

Table 10.1 Influences of soil biota on soil processes in ecosystems

	Nutrient cycling	*Soil structure*
Microflora	Catabolise organic matter Mineralise and immobilise nutrients	Produce organic compounds that bind aggregates Hyphae entangle particles onto aggregates
Microfauna	Regulate bacterial and fungal populations Alter nutrient turnover	May affect aggregate structure by interacting with microflora
Mesofauna	Regulate fungal and microfaunal populations Alter nutrient turnover Fragment plant residues	Produce fecal pellets Create biopores Promote humification
Macrofauna	Fragment plant residues Stimulate microbial activity	Mix organic and mineral particles Redistribute organic matter and microorganisms Create biopores Promote humification Produce fecal pellets

Source: Hendrix, Crossley, Blair and Coleman, 1990

The manner in which these populations are affected by various farming practices determines how well they contribute to the sustainability and productive capacity of the soil. Soil biota are affected by the physical changes in their habitat — changes in moisture, aeration, temperature and organic matter and by chemical changes, especially those caused by toxic chemicals and other pollutants. Thus the type and frequency of soil tillage and the application of insecticides, herbicides and fungicides all affect the complicated interactions between the various soil populations. In general terms, active and healthy soil biota need sufficient organic matter, aeration and drainage, and an absence of toxic levels of biocides in the soil profile. As a group, earthworms represent a major element of the biota responsible for improving soil structure. They are often used by organic farmers as one of the most reliable indicators of soil condition since their abundance reflects high organic matter, good aeration and structure, and an absence of lethal toxin levels. Such bio-indicators are gaining importance for monitoring sustainability under various cropping systems.

There have been fruitful attempts to manage soil biota, especially by farmers who appreciate their central role in fertility maintenance. For example, inoculating seeds with rhizobia (nitrogen-fixing bacteria) has been standard practice for half a century due to its direct economic benefit to soil fertility. More recently inoculation with mycorrhizae (root-penetrating fungi) has significantly improved nutrient availability for crops. Lynch (1987) has reported on the process whereby the sugars released by the fungal decomposition of wheat straw can be used by non-symbiotic (free-living) nitrogen-fixing bacteria. Earthworms have been successfully managed, first by introducing adapted species to wormless soil, then by adding organic matter and eliminating toxic chemicals. In most systems the biota are managed indirectly by regulating their supplies of energy, minerals, moisture, temperature and oxygen, and by tillage practices, crop residue management and the careful use of chemicals.

10.7 Organic farming — is it synonymous with sustainable farming?

We have already referred to the high inputs in modern Australian farming and the search for ways of reducing production costs. The actual expenditure on each item in the production process varies according to crop and animal types, but for the farm sector as a whole the 1987–1991 breakdown costs were as given in Table 10.2.

Table 10.2 Real expenditure on inputs by the farm sector in 1990–1991 dollars

	Annual average 1987–88 to 1989–90	
	$M	*%*
Fuel	1,070	5.1
Fertiliser	1,112	5.3
Chemicals	884	4.2
Seed and fodder	2,115	10.2
Marketing	2,196	10.6
Repairs and maintenance	1,753	8.4
Other *	2,879	13.8
Wages	2,585	12.4
Interest paid	2,839	13.6
Capital expenditure	3,363	16.2
Total	20,796	100.0

* Large increase in 1990-91 due to a sharp rise in the wool tax.

Source: Australian Bureau of Statistics, 1990, 1991

The need to build up and maintain organic matter has been recognised for several hundred years, but today's connotation of the term 'organic' has more to do with an absence of toxic chemicals than with organic matter as such. As mentioned in section 10.1, there is now a tendency to equate 'organic' with 'healthy' and to imply that organic farming is sustainable farming — some would claim the only real sustainable agricultural system.

Surveys show that organic farmers apply a wide range of practices and that many of them use organic methods for pragmatic reasons (e.g. low inputs) rather than for product purity or other altruistic reasons. A useful comparison was made in a survey of 77 organic and 32 conventional farmers by Massey University in New Zealand (Bakker, Gordon and Robertson, 1991), which found that while most conventional farmers had farmed all their lives, most organic farmers came from diverse non-farming backgrounds and had a very wide range of values and approaches to farming. Organic farms were generally only a quarter of the size (average 123 ha) of conventional farms (average 410 ha). Conventional farms tended to concentrate on traditional products such as mutton, beef and wool, while organic farms displayed a greater diversity and often included fruit and vegetables as well as animal products. The Massey Study found that the cropping and grazing systems used on organic farms were not significantly better at soil conservation, despite their use of compost, fish-based fertiliser and rock phosphate. Conventional farmers limited their fertiliser to superphosphate, potassic super, longlife super and urea. Both groups used agricultural lime. Of the problems encountered, organic farmers ranked internal parasites as their worst, while conventional farmers regarded parasites and flystrike as equally serious. Facial eczema and grass staggers were common on conventional farms but were not mentioned by any organic farmers. Of the weed problems, thistles were much more serious on organic farms. More than 25% of organic farmers processed some of their produce on the farm and usually sold it under the organic label. Approximately 40% of all organic produce was sold to conventional outlets. The farm provided only 54% of organic farmers' income, but 94% of conventional farmers' income. For the labour component there were no differences between the two groups and labour requirements remained the same for farmers who changed to organic farming. There were no significant changes in yields or income per hectare in the first or subsequent years for those who changed to organic farming.

So far, studies comparing organic and conventional farming in Australia have been inconclusive, mainly because of the wide range of farming activities in both groups. Virtually all producers are agreed on the need to reduce soil losses, minimise the use of chemicals and improve soil structure through minimum tillage. The problems arise where the export of nutrients in crops must be counteracted by some replacement strategy. For instance, an average Western Australian total state wheat crop (1.88%N, 0.23%P, 0.35%K) may remove:

95,880 tonnes of nitrogen
11,730 tonnes of phosphorus
18,355 tonnes of potassium.

Most of the nitrogen loss can be made good by using legumes in the wheat rotation. But what of the other major nutrients? It is difficult to see how organic forms of phosphorus and potash can be made available on the scale required for broadacre agriculture. Rock phosphate can make a useful contribution, but the danger of nutrient rundown in the absence of commercial fertiliser is a real challenge to large-scale organic grain production at present. This is not to suggest that the dependence on organic fertiliser is not appropriate to small fruit and vegetable farms where the addition of sufficient organic nutrients is practical. Organic farming has of course always been the basis of Asian subsistence farming where rice and rice straw are recycled virtually 100% by returning crop, animal and human wastes in traditional closed systems.

Sustainable agriculture in Australia is likely to draw on the principles of organic farming in its less extreme form, but is also likely to draw on modern production methods adapted to soil protection. Part of the attraction of organic farming is apparently its image of being different, down-to-earth, cheap and even spiritual in the sense that its proponents' commitment to organics are based as much on faith as on demonstrable scientific evidence.

An unfortunate consequence of the emergence of organic farming in Australia has been that competition for producer acceptance has polarised the organic and 'scientific' agricultural organisations. To some extent this 'us and them' situation was predictable — after all, the established reputations of researchers and advisers were challenged when so many amateurs (organics) claimed that the professionals (scientists) were recommending unsustainable systems. While this claim is partly correct, the evidence that the proposed organic replacement system was a permanent and economically viable alternative was unconvincing. Comparisons and the search for evidence to validate organic systems are continuing. The work of Wynen and Edwards (1990) and others requires expansion if the claims made by the Biodynamic Farmers or the Biological Farmers' Association are sound enough to justify a major trend toward broadacre organic farming.

Meanwhile, the organic lobby has benefited many conventional farmers by making them check and recheck the sustainability of present recommendations; this trend is evident in the growing numbers of departmental workshops such as Organic Farming in Field Crop Production (QDPI, 1990). As for acceptance of organic farming in Australia, Hassall and Associates (1990) found that there were 1,513 commercial organic farmers in 1989, which is less than 1% of all Australian farmers. The area estimated to be under organic farming at that time was between 0.1% and 0.2% of the nation's cropland. The survey indicated that the number of farmers taking up organic farming during the 1980s had grown by about 22% each year. The majority

(71%) of organic producers were involved in horticulture, with smaller proportions producing livestock (15%) and broadacre crop/animal production (14%); this latter group occupied 68% of all organically farmed land. The 71% of organic horticulturists made up only 10% of the total organically farmed area. At the time of the survey the domestic sales of organic products were estimated at $35–$45M, excluding private farm sales which may account for 30% of the total organic sales; and organic foods constituted only 0.2% of total domestic food sales in 1990.

As mentioned earlier, the extent to which organic farming meets the needs of overall sustainable agriculture cannot be precisely determined because of the variation in what constitutes organic or biological farming. A useful way of looking at organic farming is to consider all its possible techniques, from the 'primitive' (simple) to the 'purist' (complex). One of the most useful analyses of the elements of organic farming systems developed in Queensland (QDPI, 1990), considers the following topics:

1 NUTRIENTS FROM ORGANIC MATTER

The extent to which humus affects plant growth has not been fully determined. Plants can be successfully grown in a hydroponic medium based only on inorganic plant nutrients. In many crop systems plant nutrients become more available as soil organic matter increases, but since the benefit that organic matter confers on soil may be eventually lost, organic matter is regarded not so much as a source of nutrients but as a soil ameliorant that enhances infiltration, aeration and moisture-holding capacity.

2 NITROGEN FIXATION BY LEGUMES

The factors affecting leguminous N fixation are mostly well understood. Record levels of fixation such as 700 kg N/hayr by white clover in New Zealand far exceed the 75 kg N/ha after a year of lucerne or 45 kg after medics on the Darling Downs. In southern states the N range is 34–155 kg for medics and subterranean clover. Roberts (1960) demonstrated that when grown with white clover and lucerne, perennial ryegrass and fescue pastures in South Africa had a 3% higher protein content than when grown alone. Since minimum tillage systems tie up N but improve soil moisture, they must be evaluated to determine whether N is limiting plant growth.

3 PEST CONTROL IN ORGANIC FARMING

Apart from synthetic chemical pesticides, a range of cultural and biological methods can suppress insect pests. These include resistant crop varieties, crop rotation, cultivation, the timing of planting, planting machinery (press wheels), insect predators, natural repellents and natural insecticides (pyrethrum). In southeast Queensland, for example, cotton pest insects are susceptible to 57 predator and parasite insects, and to four insect diseases. The main obstacle to changing from chemical to biological insect control is the lack of information on how to integrate a range of biological methods with present cropping systems, although two natural bacterial strains that suppress Heliothis in cotton are already commercially available. Present commercial interests, however, encourage the expansion of chemical control, and if this method is curtailed growers may have to accept lower yields, albeit at lower costs. The natural control of pests by predatory insects has sometimes been destroyed by chemicals. For example, in the 1950s Dieldrin was used against the soldier fly in sugar cane, but this also eliminated its predators. The soldier fly returned, unchecked, as a 'secondary pest induced by organochlorine pesticides' (Robertson, 1990). The biological control of most pests remains unstudied.

4 WEED CONTROL IN ORGANIC FARMING

Organic (non-chemical) weed control methods are already widely used by farmers. Farmers converting to organic farming may have to reduce herbicides in stages rather than abruptly. Comellack (1988) estimates that weeds incur a loss of $1,870M in Australian crops annually. Farmers spend $1,320M annually on weed control: 59% of this is spent on mechanical control (cultivation), 20% on purchasing herbicides and 15% on applying them. Since crop rotation and dense planting are the main non-chemical methods of control, a change to organic farming would not be difficult in many regions. The adoption of organic weed control depends not on new technology but on new attitudes. The growing demand for pesticide-free wheat on the European market at up to $800/t has prompted wheat marketing authorities to encourage organic production methods. Conacher and Conacher's (1983) surveys of Australian organic farmers showed that weed and pest control were two of their major problems — the others were the lack of advice from official sources and the lack of reasonably priced organic materials, notably fertilisers. An analysis of research done over the past 50 years indicates that less than 2% of published research on weeds dealt with biological control.

5 CROP DISEASES IN ORGANIC FARMING

Disease control options are in two categories: chemicals and cultural practices. For wheat and barley, seed dressing with fungicides greatly reduces disease risk. Careful seed cleaning and the use of disease-resistant varieties could eliminate seed dressing, but some diseases are best controlled by changing rotations, using a wider diversity of crops, reducing crop residues (which host diseases such as leaf spot), and incorporating grazing into the production systems. In general terms organic farming may reduce or increase plant diseases, depending on the crops grown and the pathogens present.

6 SOIL MOISTURE AND ORGANIC FARMING

Organic matter regulates soil moisture by improving soil structure and the water-holding capacity of the cultivated layer. It increases infiltration, reduces waterlogging and promotes root growth. Its ability to increase the water retention of lighter (sandier) soils in temperate climates is well documented. It is not so well documented for subtropical climates, though it is known that at higher temperatures organic matter tends to oxidise (and disappear) much more rapidly than in cooler climates. Annual cultivation by heavy machinery compacts the soil, causes structural breakdown and reduces organic matter in fallows. Queensland trials (Bridge, 1990) indicate that higher levels of organic matter can be maintained in the soil by reducing tillage. This encourages organisms such as worms, which form 'macropores' in the soil, and thereby improve rain infiltration. These trials link severe compaction in non-smelling soils with low organic matter levels. The pasture phase (ley) is the best time to increase organic matter and improve surface soil structure. It can greatly increase infiltration on degraded non-smelling soils and in some cases, but not all cases, it can increase available soil moisture capacity. In some cases a vigorous pasture such as kikuyu is the only way to restore good soil/water relations.

7 SOIL CONSERVATION IN ORGANIC FARMING

Soil losses can be controlled by mechanical (earthworks) methods and biological (plant-based) methods. Contour banks have long been well accepted by producers but minimum tillage and stubble retention are more complex to integrate with the total farming system. Trials have shown that a cover of green crop growth or crop residue can reduce soil movement (erosion) to 20% of the bare soil rate, on condition that a 30% surface cover of crop material is in place during the erosive period (in this Queensland case, December to February). Experimental results show that various forms of conservation tillage, combined with contour banks (preferably sown-over broadbased banks) increase soil stability (measured by soil loss), by increasing infiltration after breaking the erosive power of raindrops at the soil surface. The PERFECT model (Littleboy et al., 1989) makes it possible to predict erosion rates with simulated cropping systems. In the eastern Darling Downs on deep clay soils under continuous wheat, predicted accumulated soil losses of 2,300 t/ha would reduce mean wheat yields to 50% of that on uneroded soils. The model predicts that a zero tillage system would maintain the soil for about 1,000 years, and that a stubble-incorporation system would maintain it for 150 years. If zero tillage systems continue to rely on chemical weed control, the type and rate of herbicide application may have to be modified.

8 ROLE OF LIVESTOCK IN ORGANIC FARMING

The author has described how animals maintain soil fertility elsewhere (Roberts, 1971). Animals are used to control weeds, to cycle nutrients (notably nitrogen), to produce manure for composting or direct application, and for diversifying production to diminish risk. Animal manure has to be applied at a rate of about 6 t/ha to give economic results. Rates over 12 t/ha are uneconomic since it costs about \$20/t spread in the paddock, assuming \$8/t at the source, \$5/t cartage and \$7/t for spreading. Applied at 6 t/ha, manure adds 72 kg N, 48 kg P and 144 kg K per hectare if their content in the manure is approximately 1.5% N, 0.8% P and 2.3% K (Thompson, 1990). The incorporation of animals complicates management on crop farms considerably, especially when expensive fencing and watering is required. The problems and potential of animal integration have been analysed by Honnef (1986) who compared a conventional grain farm with an organic farm that included animals in the Dalby district of Queensland. Unexpectedly, this study showed that there were more operations on the conventional grain farm (no animals) — 36 operations compared to 27 on the animal/grain farm. Adjusted gross incomes were $178/ha/yr for the grain farm and $250/ha/yr for the animal/grain farm.

9 WHOLE FARM PLANNING AND ORGANIC FARMING

In a state survey of what conventional farmers regarded as the chief obstacles to organic farming, southeast Queensland farmers listed lack of cash reserves to change over to organic methods, insect and disease control, low soil fertility, lack of information and lack of incentives (Vance, 1990). Equally important were the benefits these farmers expected to gain from changing to organic farming, and they ranked them as follows: (1) lower input costs, (2) premium prices for produce, (3) increased (long-term) yields, and (4) improved soils. In the context of whole farm planning (now termed Property Management Planning (PMP)) the risks of changing systems discourage farmers from adopting new practices unless these can be introduced on a small scale or step by step so that the production system (and even the farmer's lifestyle) can be modified without disrupting income and security.

10 ECONOMICS OF ORGANIC FARMING

Much of the attraction of organic farming is the promise of lower costs, higher commodity prices and the reduction of chemical hazards, especially those that threaten farmer family health. One way of moving toward organic farming is to select crops that give satisfactory yields on relatively low-fertility soils and are less prone to insect and disease damage, to seek economical alternative sources of fertiliser, and to sell on specialist markets. Gaffney (1990) maintains that the availability and labelling of organic products must improve before the domestic market for them can be expected to expand permanently. In addition, non-chemical storage must be expanded. Gaffney uses the example of a 1,200 ha Western Downs farm with 400 ha pasture to calculate a gross income of $230,000 for conventional and $208,000 for organic farming. Variable costs for organics drop from $134,000 to $108,000 and with similar fixed costs, return on capital was estimated at $26,000 for conventional and $30,000 for organic farming. With these results, the researchers concluded that organic farming warrants serious attention by those responsible for future directions in agriculture.

Chapter 11

BUILDING BLOCKS AND ACTION PLANS FOR SUSTAINABLE CROPPING SYSTEMS

This chapter explains what has to happen before Australian agriculture can move toward fully sustainable systems. We examine tillage and pest control techniques, and evaluate the complementary roles of earthworks and surface management in controlling soil erosion losses. We look closely at agro-forestry and consider how animals can be integrated with agriculture. Special attention is given to Ecologically Sustainable Development as a national program of land use planning and management. The chapter concludes by identifying the priority actions for the Landcare movement and explaining how these can be taken, given the necessary conditions.

11.1 Moving from high-input systems

We have already summed up the four biological cornerstones of sustainable cropping: (1) the reduction of soil loss, (2) the maintenance of organic matter, (3) the avoidance of nutrient loss, and (4) the minimisation of toxic accumulation (Roberts, 1992). The concept of high-input agriculture has been criticised because it leads to environmental degradation, low economic stability and overproduction. Today's sustainable agriculture is aimed at reducing chemical and energy input while still achieving long-term productivity and environmental compatibility (Poincelot, 1987).

The traditional western high-input systems are typically monocultures of annual crops that have low resistance to shocks from weather, pests and weeds because of their low biodiversity and susceptibility to single-product price structure changes. Sustainable agriculture emphasises long-term yield stability and minimal environmental impact. Maximum short yield is not a goal, but sustainability cannot be achieved simply by reducing input — new crop farming techniques are required.

In Table 11.1 Stinner and Blair (1990) sum up the basic differences between conventional cropping and what they term 'Innovative-Sustainable' cropping, and compare them according to inputs and farming operations.

Table 11.1 Comparison of ecological characteristics between conventional and innovative systems

	Conventional	*Innovative-Sustainable*
Fossil fuel energy	High	Low
Labour/management	Low?	High?; more complex
Fertiliser	Inorganic	Organic
Tillage (soil disturbance)	Lower	Lower
Crop diversity	Low	High
Life history characteristics	Annual	Mixed/more permanent
Pests	Less stable/chemical control	More stable/biological and cultural control
Nutrient cycling	Open/pulsed — emphasis on physical/chemical control	Closed — emphasis on biological control
Animal integration	Low	High
Importance of decomposer process	Low	High

Source: Stinner and Blair, 1990

It is widely believed that high inputs into agroecosystems tend to mask the important ecological processes and trends which could act as an early warning system of the breakdown of biological productivity within the soil. Converting from simpler but more expensive traditional cropping systems usually requires a higher level of management and a keener sense of observation. Sustainable systems also require expertise in the culture and marketing of a range of products that each have their own quality standards.

11.2 Cropping, tillage and pest control systems

The essence of sustainable cropping is the replacement of monoculture (one type of crop only) to a range of crops grown in rotation. The term *multiple cropping* includes many different systems, all characterised by a diversity of crops. Including legume crops to supply natural fertility to the system through *symbiotic nitrogen fixation* is vital to multiple cropping. Broadening the diversity of plant species and plant structures (height, rooting depth, density) makes better use of soil moisture and reduces pest damage. Greater crop diversity mimics natural plant communities and enhances the biological control of pests and diseases. Total crop failure is minimised by multiple cropping although the financial gains from having maximum acreage under the most profitable crop are forgone in favour of risk-spreading.

There are significant constraints to the general adoption of multiple cropping, notably the dominance of cultivars bred for high yields from high-input agricultural systems. Crop characteristics such as competitiveness, shade tolerance, moisture use efficiency and plant shape will need more attention in future plant breeding programs if crop characteristics are to match system requirements.

Apart from more complex management and the delay in breeding crops suited to diverse low-input systems, conversion to multiple cropping is currently discouraged by government policy which offers differential incentives to producers who grow single crops marketed by powerful national control boards. In other countries such as the United States, the subsidisation of particular products, notably grain crops, has exerted a dominant influence on farmers' inclination to move to multiple cropping.

After making the basic decision to adopt multiple cropping as the foundation of a more permanent production system, the next most important decision concerns tillage methods. If we define *conservation tillage* or *minimum tillage* as any soil preparation that leaves at least 30% of crop residues on the surface, a wide variation in mechanical procedures is available to operators wishing to adapt their operations to particular soil types, crop residues and erosion hazards.

Most sustainable crop systems include tillage systems which minimise soil disturbance, over-aeration, structure breakdown, compaction and loss of organic matter. These systems replace the traditional mouldboard and disc ploughs with tined implements that do not invert the plough layer of soil. The overall effect of such conservation tillage is to reduce runoff and thus increase soil moisture for crop production while aiding organic matter buildup. As runoff decreases, so does soil erosion.

Conservation tillage has been adopted rapidly, if not for conservation reasons then for economical reasons since the reduced energy input required by minimum tillage significantly cuts production costs. In its most advanced form, minimum tillage becomes *zero tillage* which entirely eliminates cultivation. In such systems, crops are planted directly into the standing stubble of the previous crop — thus the alternative term, *direct drilling*. The control of weeds by mechanical methods is thus replaced by chemical methods. This inclusion of herbicides, as an essential element of conservation tillage, is at odds with farmers' goals of *chemical-free* production. However, provided the herbicides do rapidly degrade into harmless breakdown products, no problems of toxins are encountered.

The reduced soil disturbance under minimum tillage enhances the accumulation of organic matter and humus by eliminating any excessive oxidation from incorporated plant material. This in turn improves soil structure and physical stability, thus making the surface less susceptible to water erosion. The complex interactions between soil fauna, microbes, nutrient cycling and plant growth are not fully understood but appear to have the overall effect of increasing the biological regulation of soil processes. In the United States the incorporation of crop rotations with legume cover crops has meant that only very low applications of herbicides are necessary, and in some cases they have been eliminated from the system.

The third major building-block of sustainable crop systems is the elimination of high levels of pesticide applications. This is achieved by combining the first two elements of the system (multiple cropping and minimum tillage) with the strategic selection, combination and application of insecticides. This approach is particularly important in pest-prone irrigated crops such as cotton which have become regarded as environmentally damaging because of the heavy applications of insecticide formerly used on this crop. Figure 11.1 shows how the input levels of agricultural chemicals reduce as alternative approaches are adopted.

Studies of the life cycles and breeding habits of crop pests have made it possible to apply much lower but strategic doses of insecticide that achieve the same or better results attained by higher 'blanket' spraying at regular intervals in earlier years. The development of 'softer' chemicals which are less persistent and chosen for specific pest insects has markedly improved the 'environmentally friendliness' of intensive crop production.

In multicrop systems the life cycle of many crop pests is broken or at least disturbed. For example, a diversity of crops encourages a wider range of predator insects which are now becoming relied on more than ever for the *biological control* of pest insects.

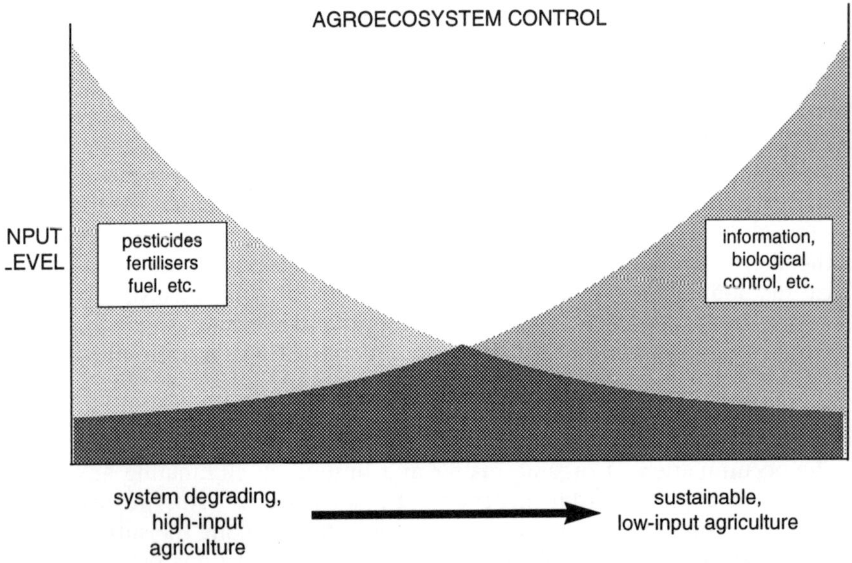

Figure 11.1 Phase-out/phase-in relations between high and low input cropping systems

Edwards, 1990

11.3 Erosion control earthworks

In earlier years the recommendations for sound cropping practice concentrated on the physical structures required for soil conservation. These include contour banks, stormwater drains, grassed waterways and pasture furrows, all of which are designed to slow runoff water and thus reduce soil loss. The essential features of soil conservation works have varied little since Bennett's (1939) original proposals although design specifications have been adapted to regional soil types and rainfall intensities as shown by Hudson (1976). An overview of the integration of earthworks and other conservation practices is given by Roberts (1984b) for subtropical conditions in Australia.

In contrast to this heavy reliance on earthworks for sound conservation farming in earlier years, Australian agricultural literature now reflects an important trend toward a more integrated conservation program that depends on a combination of crop choice, tillage methods, rotations, strip cropping and earthworks. The contribution of contour banks to agroecosystem stability depends largely on the slope of the cultivated land. On slopes of less than about 1.5%, soil conservation can usually be more readily achieved by strip cropping — planting alternate strips of summer and winter crops at right angles to the direction of surface runoff flow.

In evaluating earthworks and other conservation methods, Hudson (1992) compares the function and thus choice of a wide range of earthworks and other physical runoff barriers. He emphasises that the agronomic support systems associated with these works require further research before the full value of the physical barriers can be properly compared. He groups the barriers according to their goals — soil management, water management and crop management — as shown in Table 11.2.

After 40 years of soil conservation experience worldwide, Hudson concludes that the key to getting programs in place is the involvement and willingness of the landholders. This involves several changes in soil and water management priorities:

- from projects to cure erosion to programs to prevent it
- from thinking 'soil conservation' to increased productivity through 'land husbandry'
- from earth-moving soil conservation to better farming
- from quick-fix projects to long-term programs
- from top-down plans to people-led development.

11.4 The benefits of tree/crop systems

Tree planting has become synonymous with rural conservation in Australia. The link between salinity and tree clearing has led to a public image of trees as the instrument for rehabilitating rural land, irrespective of the role of trees in particular landscapes. Programs such as

Table 11.2 Terraces and other cross-slope barriers for different purposes

Main Objective	*Function*	*Type of terrace or barrier*
Soil management	To modify soil slope.	Bench terraces.
	To slowly reduce soil slope.	Progressive terracing*
	To contain erosion with low inputs.	Ladder terraces
	To contain erosion with minimal earth-moving on steep slopes.	Trash lines*; step terraces*; hillside ditches*; intermittent terraces*
Water management	To multiply rainfall benefit.	Conservation bench terraces; run-on level terraces
	To catch and hold all the runoff.	Absorption ridges
	To absorb some runoff with	Contour furrows
	emergency overflow.	Contour bunds
	To control unavoidable runoff.	Graded channel terraces
	To control reduced runoff.	Ridging; tied-ridging*;
	To reduce the velocity of runoff and promote infiltration.	Strip cropping*; grass strips*; permeable barriers*
Crop management	To provide level areas on steep slopes, or to ease cultivation according to whether by hand, ox, or machine.	Step terraces*; hillside ditches*; orchard terraces*; platforms
	To ease harvesting according to whether the crop is heavy, damageable, harvested regularly or seasonally.	Footpaths and farm tracks* associated with orchard terraces or hillside ditches
	Drainage for crops suffering from wet feet.	Ridges on 2% grade for tobacco; up and downslope beds for yams; small open drains up to 15% for teff

* Practices with lower labour requirement than conventional terraces.
Source: Hudson, 1992

One Billion Trees and Greening Australia reinforce the impression that trees in degraded catchments have healing power.

The other side of the tree debate is the hard evidence that well selected trees, in locations to which they're adapted, can contribute substantially to land improvement and stabilisation. The productive integration of trees with cropping systems is termed 'Agroforestry'. It is an ancient practice and Bishop (1983) suggests that agroforestry was probably the original type of farming in forested regions.

Today agroforestry is being researched as one of the most promising forms of sustainable agriculture. It is compatible with both annual and perennial crops and can incorporate animal production into the system. As a system, agroforestry can achieve objectives that pure cropping systems cannot reach. Used correctly, the inclusion of trees in the system can improve nutrient cycling, ameliorate microclimate, and reduce wind erosion and water runoff. Trees can increase the biodiversity of pest predators and beneficial insects, and expand the range of foods and other products yielded by cropping systems. This range of products includes food for humans (fruit, nuts), food for

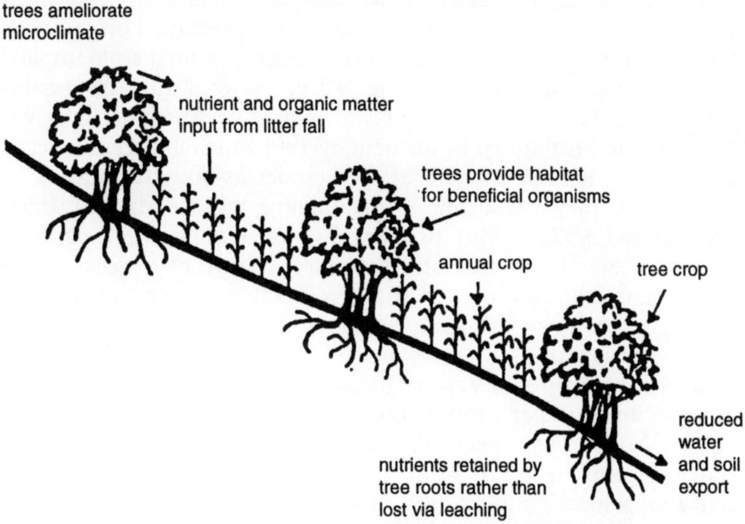

Figure 11.2 Diagrammatic structure and processes in agroforestry systems
Stinner and Blair, 1990

livestock (fodder, seed, pods), fuelwood, timber and food for wildlife and pest predators. The structure and ecological function of agroforestry systems is shown in Figure 11.2.

Agroforestry may concentrate on using the trees to improve the micro-environment of the crops sown between the rows or strips of trees. In such cases, the trees act as windbreaks and improve both temperature and atmospheric moisture for crop growth. In other cases, the trees are regarded as crop plants in their own right and may yield at least the same income as the interplanted crops. Apart from integrated crop/tree systems, trees in belts, patches or blocks benefit landscapes rendered virtually treeless by the expansion of grain and pasture production in developed countries. In a third form of agroforestry, a succession of increasingly larger trees follow each other in a progressive ecological series which mimics natural plant succession. This approach has been the basis for certain tropical systems for many centuries but has also been successfully used as a modern alternative to monoculture (Hart, 1986). It is claimed that these successional systems yield up to 31% more protein and up to 54% more calories than monocultures in the same region.

Trees have a special role in degraded catchments where afforestation of the upper slopes can stabilise the soil and the runoff patterns in a way other systems cannot. While this is particularly the case in tropical regions, agroforestry could well be the key to what has been termed a 'reverse Green Revolution', in which traditional agroforestry replaces broadacre monocultures.

In attempting to predict the advantages of including trees in the production system, it is useful to examine the presumed original benefits of removing trees from Australian landscapes on a scale unparalleled in human history. The cycle of tree removal and rising saline groundwater tables is well known in temperate Australia. Less well known is the breakdown in nutrient cycling which apparently occurs in tropical Australia where trees which act as 'mineral pumps' on infertile soils play a vital role in maintaining pasture plant nutrition. Research by CSIRO (1989) indicates that tree removal alone confers a negligible benefit on grass growth under a range of tropical (North Queensland) conditions, and that irrespective of tree numbers the carrying capacity of the pastures depends on the extent to which legumes are sown.

In Australia agroforestry has usually been construed as combining cropping with timber production, rather than recognising the other benefits from trees. Special attention has been given to the crop/timber systems in Victoria and in the southwest region of Western Australia, where *Pinus radiata* has been extensively evaluated. In pasture/pine combinations, the spacing, pruning and age of the trees when harvested were important to obtaining good pasture yield as well as acceptable volumes of timber. Much of the debate over agroforestry concerns how shading and tree roots affect the yield of associated pasture or grain crops. In countries such as New Zealand the main purpose of *Pinus radiata* is to stabilise soil against land slip and erosion in steep country. In Australia, apart from pines for timber production, a range of indigenous trees — especially *Eucalyptus* and *Acacias* but also *Casuarinas* and *Angophoras* — have been planted on a vast scale for salinity control. The details of these planting techniques are given by the writer elsewhere (Roberts, 1992); at this point we consider how incorporating trees with crops contributes to sustainable agroforestry systems.

At farm level, the role of trees in whole farm planning has been given special attention by the Landcare movement since its inception. Today demonstration farms can be visited in every state where a range of different forms of agroforestry can be inspected. For example, the Stockman Stud at Kempton in the Tasmanian midlands (Richardson, 1994) — a 1070 ha property that is used almost entirely for wool production except for 5% of the land under crops. This open rolling country was extensively cleared from the 1870s to the 1940s. Today a farm plan controls water erosion by integrating pastures, trees and earthworks by using shelter belts and 45 km of new fencing to protect woodlots and by regenerating erosion areas and remnants of natural scrub. New pastures (300 ha), salt-tolerant grasses (30 ha), lucerne (25 ha) and tagasaste fodder trees (4 ha) all contribute to a dense protective soil cover which has stabilised the farm against erosion and salinity. The owners, Gerard and Elizabeth McShane, have planted 24,000 trees. Of these 10,500 are in shelter belts and habitat areas, 4,500 are for erosion and runoff control, 4,200 are for fodder protection, 3,000 combat saline discharge and 200 tagasaste trees are used as a seed source. Natural regeneration sites of 10, 20 and 50 ha were reinforced with additional native species, and 15 km strips of native trees were direct-seeded to combat wind erosion.

This type of integrated agroforestry system is becoming widely adopted not only in Tasmania but in other states, especially Victoria where Farm Tree Programs, which pre-date Landcare, have been active for nearly 20 years. In Western Australia, Greening Australia, Alcoa and state departments have made significant progress in getting agroforestry accepted in the state that once prided itself on clearing 'One Million Acres per Year'.

11.5 Integrating animals with agroforestry

As grain farming became more specialised and more profitable, agriculture in many regions of Australia moved away from integrating trees and animals with the farming system. Today many of the ills of intensive cropping are thought to be caused by the increasing lack of biodiversity (range of species) in the system. The development of insecticide-resistance in pests, the reduction of soil organic matter and the increase in plant diseases are some of the symptoms of this 'unnatural' monoculture. Many alternative farmers propose returning trees and animals to traditional grain production systems. Whether these approaches are termed 'organic' or 'biological' or 'Permaculture', their basis is the same: a return to a diverse, balanced plant and animal population by integrating annuals and perennial crops, animals and trees as cooperative elements of a *symbiosis* (a 'living together') which enhances natural control mechanisms in the systems.

The role of the animals in the system varies with the crop/pasture rotation, but in general, animals serve to cycle nutrients by returning dung and urine to the soil, and may also serve to control weeds by grazing them and preventing them from setting seed. Management becomes complex when planted tree seedlings require protection from animals using the same land, so well-planned fencing or tree-guards are an essential tool.

The unexpectedly large response of crops to manure applications indicates that biological factors other than the nutrient content of manure are operating to give yields considerably greater than the manure's chemical analysis would lead one to expect. The classical fertiliser trials at the Rothamsted Experimental Station in England demonstrated over a period of 117 years that when equal amounts of phosphorus were applied as superphosphate and manure, manure did better at raising soil phosphate levels. It has also been shown (Brogen, 1981) that after 25 years of applying manure, green manure (green legumes ploughed in) and mineral fertiliser, both types of manure led to a 200–300% greater increase in nitrogen in the soil humus than achieved with mineral fertiliser. Such results strengthen the case for integrated animal/crop systems based on lower inputs of synthetic chemicals.

11.6 Current ESD recommendations for agriculture

Several federal and state conservation programs reflect the growing awareness of land degradation as a major issue in attaining ecologically sustainable development. Strategies such as the National Conservation Strategy and the Plan for the Decade of Landcare offer a useful framework on which to expand sustainable land management. Currently the most comprehensive national set of proposals to encourage sound farming practices is the list of 27 recommendations made by the ESD Agriculture Working Group, presented here in abridged form:

A RESOURCE MANAGEMENT — GENERAL

Rec 1: That federal and state governments support ESD in agriculture through (a) land management that incorporates sound principles of environmental protection and of economics, (b) community self-help groups and (c) integrated farm planning.

B LAND MANAGEMENT

Rec 2: That the concept of self-help groups are expanded so that they can attend to any aspect of resource management; that further technical support is given to these groups; that additional financial resources are sought for them and that existing programs are coordinated to ensure that sustainability objectives are achieved.

Rec 3: That land management programs are further coordinated by integrated catchment management in all states; that programs focus on strategic and participatory action plans for whole catchments, considering environmental, economic and social objectives.

Rec 4: That whole-farm planning is adopted as the basis for farm management, including physical, resource and economic planning combined with sociological and business issues; that a specific program is developed to promote whole-farm planning to financial institutions, state departments, rural adjustment authorities and educational institutions; that 50% of the cost of plans that meet ESD objectives is paid through tax incentives and grants; and that government agencies are given coordinated resources to help them support farm planning.

Rec 5: That the procedures for consultation and approval of commercial agricultural projects which could degrade the environment are reviewed; that the efficiency and consistency of state and Commonwealth approval procedures are improved.

Rec 6: That the terms and conditions of Crown leases are reviewed to encourage ecologically appropriate management based on property plans.

Rec 7: That pest control legislation is reviewed to make it appropriate for private and public land; that landholders accept responsibility for pest control; that government make resources available for regional control programs where necessary; that integrated pest management research is expanded; that special measures are developed to control native and feral pest animals in rangeland areas; and that control measures for kangaroos are improved, including the evaluation of a nationally regulated commercial kangaroo industry.

Rec 8: That landholders are encouraged to protect remnant vegetation by educating them, by strict clearing guidelines and sanctions, and by controlling all herbivores affecting remnant vegetation.

Rec 9: That governments support the use of private land for conservation purposes and wildlife reserves; that capital and management costs of private reserves are supported and compensation paid for resumed land or land where production is restricted for conservation purposes.

Rec 10: That financial support is given for water harvesting and distribution, conservation-related fencing, minimum tillage machinery and revegetation.

C WATER RESOURCES

Rec 11: That an environmentally oriented water pricing system is introduced allowing entitlements to be transferred between users and between states; that environmental requirements are considered when governments allocate water; that systems of allocation are reviewed to give users more control over efficiency; and that guidelines are developed to prevent water pollution from agriculture.

D AGRICULTURAL CHEMICALS

Rec 12: That the Select Committee on Agricultural Chemicals' recommendations on spraying, containers and training are acted on.

Rec 13: That the National Residue Survey is expanded to meet future needs and that the Market Basket Survey is improved and updated to include rankings of foods on health risk.

Rec 14: That research on and monitoring of agricultural chemicals provide more information to rural doctors, give more details on health hazards, examine the long-term effects of herbicides, monitor effects on wildlife, minimise the use of chemicals and develop cheap tests for residues.

Rec 15: That specific chemicals are reduced so alternatives can be encouraged and market standards met.

E EDUCATION AND ADVISORY SERVICES

Rec 16: That education courses at all levels clearly identify ESD principles and integrate them with teacher training; that government promptly provides urgently required extra funding for ESD course development.

Rec 17: That group leader training is funded through appropriate programs to educate community group leaders in all aspects of self-help action, with special attention to women trainees.

Rec 18: That financial support for the expenses incurred by community group organisers is paid by government.

Rec 19: That resources are made available to develop and promote decision-support packages for ESD in agriculture.

F AGRICULTURAL RESEARCH

Rec 20: That research into the sustainability of major agro-ecosystems is supported by commodity-based bodies across different industries.

Rec 21: That practical interdisciplinary research programs into ESD are developed incorporating landholders' opinions with research planning.

G ORGANISATIONAL STRUCTURES

Rec 22: That all government-supported bodies in agriculture and land management ensure that ESD principles are embodied in their goals.

Rec 23: That the activities of the following major resource bodies are integrated as a matter of urgency: the Agricultural Council, the Soil Conservation Council, the Water Resources Council and the Environment and Conservation Council; that such integration focuses on making a consistent agricultural resource policy.

Rec 24: That submission and approval procedures for land management projects by self-help groups and researchers are coordinated and consistent so that only single applications are necessary.

H CLIMATIC CHANGE

Rec 25: That measures to limit agriculture's output of greenhouse gases are reviewed as part of a national action plan to alleviate global warming.

Rec 26: That the following are encouraged, to reduce gas emissions and achieve ESD: improved animal waste management, reduced burning of biomass, increased efficiency of energy use, reduced tree clearing and increased tree planting, increased trash retention and minimum tillage.

I MONITORING

Rec 27: That all governments publish progress reviews of ESD achievements in agriculture every five years; that a national monitoring system for environmental and economic aspects is set up to assess changes in resources and productivity.

11.7 Working Group's proposals

Section 11.6 dealt with the ESD Agricultural Working Group's 27 recommendations which covered the whole range of national issues identified by the Group after receiving submissions from all sectors of rural industry. A useful way of clarifying the relationships between all the recommendations is to rearrange them according to the priority ranking given to each one by the Working Group (Table 11.3).

Table 11.3 Priorities of ESD Working Group on Agriculture

Priority	*Recommendation*
1 Very High	• Extend Landcare activities. • Improve Landcare delivery programs. • Meet 50% of costs of preparing property plans. • Make resources readily available to assist property planning. • Develop strategic plans for managing pest animals and weeds. • Improve state procedures to provide sanctions for vegetation clearance. • Extend direct financial incentives for farm land improvements. • Complete water pricing guidelines. • Approve transferability of water entitlements. • All governments to act on Senate Report recommendations on use of chemicals. • Review national education curricula to include ESD. • Provide funding for ESD-oriented course development at all levels. • Training of group facilitators. • Financial support for group organisers. • Integration of responsibilities of all Ministerial Councils on resource management. • Streamline group project approval procedures.
2 High	• Extend resource management support through extension in Landcare. • Strengthen ICM programs. • Promote benefits of property planning. • Improve property planning approval procedures. • Improve Crown lease conditions. • Develop kangaroo management strategy. • Support nature conservation provisions on private land. • Develop better water quality guidelines. • Make more information available on toxicology of chemicals and pollutants. • Gain more information on health hazards from agricultural chemicals. • Set targets for acceptable levels of chemicals in crops and food. • Support further development of decision-support packages. • Promote adoption of ESD principles by research organisations in research planning and fund allocation.
3 Medium	• 8 remaining recommendations as listed by Working Group.

Source: WGSA, 1991

No doubt a different set of high and very high priorities would be arrived at if the list was made by Landcare Groups themselves, although general similarities of priorities could be expected. Note that the very high priorities listed above include categories of direct and indirect assistance to landholders, plus attention to many aspects of policy and procedures which affect landholders' actions. Note also that virtually all the very high priority actions are either Landcare Group activities or associated training and education activities supporting Landcare. The exceptions are water pricing and licensing, although in irrigation areas plagued by salinity such policies are fundamental to improved water management. Individuals and groups in different agro-ecological regions will probably consider that this list does not meet their particular needs, but if government acts on these recommendations promptly, they could greatly benefit sustainable farming and grazing methods.

While committees with a broad range of objectives, such as the ESD Working Group, may rank their recommendations from a national problem-solving point of view, research bodies may rank their priorities somewhat differently. For example, Australia's premier research organisation, the CSIRO, has used two prime criteria in ranking their national research priorities: (1) *attractiveness* and (2) *feasibility*. This approach to allocating resources to specific fields has particular relevance to research on sustainable land use systems. The concept of 'attractiveness' includes potential economic, social and environmental benefits to the nation at large, plus Australia's potential to capture the benefits of technical progress by converting them into commercial returns. 'Feasibility' on the other hand is evaluated in terms of how well Australia's research and development capacity can handle a research area and its outcomes.

11.8 Priorities in Landcare — the tools to do the job

The evolution of the Landcare Movement represents the most potentially effective conservation program in rural Australia's history — its success depends on changed attitudes, increased knowledge and funding group projects. At the heart of Landcare's progress is the sense of 'ownership', which concerns personal and group commitment to land improvement. Campbell (1992) has suggested a useful approach to group objectives for sustainable agriculture, based on four pillars: commitment, resources, knowledge and change process. These essential elements for Landcare group progress interact with each other — they are interdependent, not independent factors. In analysing the constraints to national progress with Landcare, Campbell (1992) proposed a series of steps for improving the progress of the community groups that comprise the basic local action units. These proposals, categorised as funding, training, resource monitoring, local government and landcare evaluation, are summed up as follows:

- Funding: (a) Establish a resource economics study to examine the financial inputs into, and the real social costs of, land degradation, to determine the rational level of expenditure required to maintain Australia's natural capital. (b) Allocate a fixed minimum percentage of the gross value of agricultural production to Landcare.
- Training: Establish a National Landcare Program training initiative with an initial budget of $1.1M p.a. to improve technical and interpersonal skills.
- Monitoring: Establish an initiative for 'Natural Resource Monitoring and Land Literacy' to identify principles, guidelines

and indicators so that groups, agencies and schools can become involved in measuring changes in the condition of soil, water and vegetation.

- Local government and community organisations: (a) Initiate a National Landcare Program-funded project in cooperation with the Australian Local Government Association, to encourage local governments to employ Landcare facilitators and become involved in the training for and monitoring of sustainable land use in their areas. (b) Set up, in conjunction with local government, the National Farmers' Federation and the Australian Conservation Foundation, a program funded by the National Landcare Program to employ Landcare liaison officers for training, monitoring and land literacy programs in each region. This is aimed at encouraging these producer and conservation organisations to play a more constructive role in Landcare.
- Evaluation: Allocate 5% of the budget for Community Landcare Support to the Landcare evaluation strategy in which all aspects of the Landcare Movement can be monitored accountably, and so that the national Landcare facilitator can play a networking role in group evaluation.

Priorities in Landcare should form a subset of the priorities in ESD; note the high priority given to Landcare group activities in Table 11.3. This suggests that at a national level these local groups are seen as the basic action units where progress will be vital to the national effort to restore productivity and stability.

Many of the Landcare groups are still at an early stage of development. Their priorities probably concern group dynamics and getting their group functioning well, rather than 'end-products' such as changing management and better soil condition. The priorities of producer organisations tend to range more widely — Landcare is considered along with the other pressing matters of rural politics and financial hardship which producer organisations become involved in on behalf of their members. As an example of what producer organisations could contribute toward the drive for sustainable production, the following priorities have been proposed for the Cattlemen's Union in Queensland (Roberts, 1992):

1 They can inform themselves of the real need for changing from systems in which production is gained at the expense of our 'land capital'. Many landholders still fail to appreciate the crucial significance of land degradation, particularly in the pastoral zone.
2 They can make valuable contributions to policy-making through the expanded consultation process, on issues such as drought declaration and relief, national park acquisition and management, land clearing guidelines, feedlot regulations, and accreditation for users of agricultural and veterinary chemicals.
3 They can seriously embrace the concept of Rangeland Monitoring — recording the condition and trend of native pastures as a basis

for animal management. This program could make a major contribution toward improving the long-term carrying capacity of pastoral country (Roberts, 1990).

4 They can offer original practical ideas for a workable set of cross-compliance schemes in which landholders are given financial benefits for applying land improvement measures — for example, the Voluntary Property Amalgamation Scheme, tested by the Queensland Department of Primary Industries in the Mulgalands.
5 They can enter open discussions with the Department of Environment and Heritage to achieve more equitable and rational acquisition and management programs. The department is now moving from its acquisition phase to its management phase, and neighbours especially should contribute to local park strategies to control vermin, weeds and fire.
6 They can participate in a wide-ranging recording and analysis of the relationships between stocking rates, economic returns and pasture condition. A start has been made on such studies by the Land Use Study Centre at the University of Southern Queensland, by the New South Wales Department of Agriculture in the Western Division, and by the Queensland Department of Primary Industries.
7 They can contribute their collective experience to developing practical but ecologically safe guidelines for land clearing in each region.
8 They can pool their experience to produce a strategy for woody weed control in affected regions.
9 They can formulate and assist with safe procedures for the use of toxic chemicals.
10 They can continue to expand their input into land administration policy concerning rentals, covenants and incentives for the maintenance of crown land in good condition.
11 They can make use of computer models for land and animal management — for example, the QDPI's 'Beefman' and the CSIRO's 'Rangepack' are decision-making tools that simulate predicted conditions and compare options.
12 They can take Landcare at face value, embrace it as the most potent cross-industry force that rural producers have ever had for maintaining their resource base. If Landcare fails, it will be because there were too many onlookers on the sideline and not enough giving it a real push in the scrum.

11.9 Conditions necessary for land management

The wide-ranging catalogue of proposals listed above is based on the growing urgency for corrective action on a series of obstacles that have hindered sound land management for years. While these recommendations are well directed, it is now time to put them into practice, and to organise the people, finances and technology necessary for a nationwide program of improved land management. The complexities and sequential stages of planning for a state Landcare program indicate the multiplicity of conditions which must be met for a community-driven sustainable land use program to succeed. The following pro forma was a forerunner of the authorised Landcare Plan for Queensland (Roberts, 1990):

Objective

To implement a program of Landcare which is essentially defined as the application of sustainable land management practices which include optimal plant cover, high plant and animal production, full use of water, maintenance of diversity and minimal buildup of toxins.

These are achieved by:

- land capability planning
- sustainable cropping techniques
- sound animal production methods
- fauna and flora retention on farms
- ecologically-based park management
- integrated tree planting for specific purposes
- mine-site rehabilitation
- roadside conservation

These practices will be implemented when landholders:

- are aware of the consequences of not acting
- recognise the benefits of acting
- respond to incentives to act
- are pressured by peers to act
- face penalties for not acting
- take pride in the results of acting
- are given practical advice on how to act
- are able to test advice on a small scale
- are accepted by peers as acting correctly

These conditions for action will be met when:

- education and information provision changes attitudes
- the benefits of landcare are clearly elucidated
- incentive schemes are promulgated
- community values include land ethics
- enforcible penalties are agreed on
- reward systems and acknowledgement are operative
- practical advice is readily extended
- discreet elements of landcare are tested under local conditions
- the community recognises and acknowledges landcare achievers as role models

These changes will be achieved when:

- the school system and advisory services incorporate a philosophy of sustainable land use

- research and experience are transmitted in a persuasive manner to gain acceptance of landcare benefits
- governments operate meaningful incentive schemes
- community education embraces land ethics
- governments enact legislative penalties
- governments and communities deliver accepted rewards
- advisory services offer and promote comprehensive information
- discreet trials and demonstrations of major landcare techniques are set up in all regions
- communities identify and publicly honour good land managers

Actions necessary to meet these objectives:

- government advisers submit specific acceptable curriculum proposals to the Education Department
- advisory services develop, agree on and implement a code of practice embracing the central concept of sustainability
- advisory services develop comprehensive sets of extension material which demonstrate the benefits of landcare for each group of land users
- land use specialists submit and gain government agreement on details of an incentive scheme to foster sound land use
- education and advisory services develop and present materials through an adult education program which fosters the addition of land ethics to the community's sense of values
- government advisers develop and gain acceptance of legally enforcible penalties for permanently damaging land resources
- government, community organisations and businesses develop and offer a range of rewards for landcare achievements
- advisory services identify gaps in available practical land management advice and initiate research to remedy each deficiency in each major land use alternative
- advisory services develop integrated information collections, combining government and other sources
- advisory services and community groups identify, plan and demonstrate proven landcare techniques in each region and commence trials to compare further alternative methods
- community groups and government draw up recommended means of publicly honouring landcare achievers in each major land use, in all regions and age groups.

These suggested actions and their prerequisite conditions highlight the need for a wide array of institutional procedures and incentives to converge on the problem-solving program. Submerged in the dynamics of community-led conservation activity is the potential role of women in leading first a change in environmental attitudes and then a change in community behaviour on environmental issues in agriculture. Brown and Switzer (1991) have given a useful analysis of the role of gender in achieving sustainable development. They argue that the 50.1% of the Australian population who are female aremostly employed in the education and health sectors where they make up 64% and 76% of employees respectively. On personal environmental priorities women score consistently higher than men on the importance of pollution, nature conservation and socio-environmental issues. With the present divorce rate exceeding 25%, women also act as head of 87% of single-parent families. Women's well-developed concern for the future of their children, combined with their innate nurturing abilities and capacity for social networking, give them extra potential as leaders of the quest for sustainable agriculture.

At a practical level, ESD depends on good research, good demonstration and extension, and strong motivation in landholders. To focus and encourage ESD research, in 1990 the Federal Government established the Land and Water Resources Research and Development Corporation (LWRRDC). This important organisation will have a significant effect on both the scientific base of Australia's land use planning, and on establishing practical techniques for land improvement. The LWRRDC is making important contributions to dryland salinity control, irrigation salinity strategies, rangeland management and other spheres of vital national concern in ESD.

At the policy level, if Australia is to set the conditions and prepare the nation for ESD, several internal and external conditions are necessary:

1 The concept of intergenerational equity will have to be made acceptable by a full environmental costing of resources, including the cost of damaging such national assets.
2 To meet the conditions for preserving our unique biodiversity and the integrity of our ecosystems, significantly more attractive incentives and rewards will be required. The current lack of penalties still encourages the kind of exploitative behaviour that has no place in ESD. Such regulatory mechanisms must be combined with better awareness, gained by education, and sufficient financial incentives for land stewards acting on the community's behalf.
3 An important subset of ESD is the concept of social equity, without which a sustainable society is unlikely to survive. Its permanence depends on all social groups accepting the inherent capability and limitations of the land as they apply to all uses and all social groups.
4 Ecologically appropriate waste disposal is becoming vital for maintaining our lifestyle. If ESD is to succeed in Australia we shall have to do a much better job of handling our solid and liquid byproducts. This means attacking our high per capita consumption levels at their source, then dealing with waste and recycling far more efficiently.
5 A fundamental condition for ESD is a basic move from the values of a materialistic consumer society to a more frugal, simple, lower-input set of community values. Redefining success in terms

of quality of life rather than our standard of living in economic terms must become an important tenet of our education system.

6 Successes and failures in many policy spheres have shown that the community must be involved if changes in behaviour are to be permanent. One of the most important conditions for ESD is thus a proper consultative process which gives the people first a sense of involvement in the goals and strategies of ESD, and second, a sense of pride in 'their' achievements. As the Chinese saying goes 'When the best leaders of all have done their work the people will say "We did it ourselves"'.

7 Finally, Australia will need a global perspective on its problem-solving strategies in ESD. There is no real gain in exporting our problems by moving them offshore in the form of toxic wastes, hazardous chemical manufacture or the use of Asian rainforest timber. In addition, our foreign aid must concentrate on developing self-sufficient communities in developing countries rather than on the politically expedient 'handout' projects all too common in the past.

Chapter 12

ENERGY AND SUSTAINABLE LAND USE PRACTICES

This chapter concentrates on the energy factor in rural production systems. We consider energy consumption as a crucial element in a future energy-conserving society and evaluate the possibilities of growing our own energy on the farm. The chapter concludes by examining how energy flows through biological production systems and how the efficiency of energy transformation can be improved.

12.1 Energy consumption — the measure of overconsumption

Earlier consideration of the attributes of a sustainable society established that consumption levels of non-renewable energy eventually limit the survival of high-input production systems. Currently the per capita energy consumption levels of the western nations are several times higher than those of the less developed nations. Australia has the dubious distinction of being third on the world list of per capita energy consumers, despite our relatively warm mid-latitude climate.

Agriculture makes a significant contribution to Australia's high energy consumption because of its direct use of fossil fuels for farm operations and the energy-intensive processes it uses for agricultural products such as fertiliser. To date, Australian agriculture has done little to reduce energy use for environmental or sustainability reasons although a reduction of production inputs have been sought for direct financial reasons.

During the energy crisis of the 1970s when Middle East oil was not available, considerable attention was given to alternative (renewable) sources of energy for agriculture. Research had shown that alcohol from sugar cane could replace petrol, and that vegetable oil from sunflowers or peanuts could replace diesel oil. Feasibility studies on converting coal to oil also indicated the potential for an Australian coal conversion industry, based on Victorian brown coal. Other coals, notably in Queensland, were shown to be at least as suitable for conversion as the South African coals used in the SASOL process of oil production for the past four decades.

Apart from running internal combustion engines on alternative fuels, agricultural operations offer a wide range of opportunities to reduce energy inputs and so reduce production costs and the environmental impact of emissions. Wittwer (1975) made a special study of energy-saving technologies in food production and found opportunities for increasing energy efficiency in soil management, water management, pest control, weed control and tree management. Energy in crop and animal production includes the biological energy required for growth and temperature regulation — for example, windbreaks, shade, metabolic heat and growth energy. Wittwer (1975) suggests that the hormonal stimulation of both photosynthesis and senescence can enhance energy-use efficiency. Frost control with shelterbelts or windbreaks and mulch can also reduce energy losses.

In soil management, techniques such as minimum or zero tillage and nitrogen fixation by legumes (instead of high-energy fertilisers) can all make significant energy savings, and crops that are adapted to lower fertility conditions and more tolerant of diseases and pests can reduce the use of high-energy products. Gravity irrigation systems, low pressure drip irrigation and windbreaks to reduce evaporation losses can reduce energy for pumping and increase the efficiency of water use. The range of alternatives for controlling insect pests — crop rotations, resistant varieties, biological pest control etc. — all help to reduce the use of pesticides, and plant disease control strategies can reduce herbicide use. Multiple cropping, agroforestry and intercropping with cover crops in orchards all help to reduce energy-consuming synthetic chemicals.

As an example of home-grown fuel energy, extensive trials in South Africa during the 1970s demonstrated that if 10% of a farm's crop acreage was used to grow sunflowers for fuel-oil for diesel tractors, that acreage could produce sufficient fuel to cultivate and harvest the remaining 90% of the crop area of the farm.

12.2 Farm-grown fuel — gimmick or future essential?

While most informed farm operators acknowledge that fossil fuel supplies will eventually come to an end, the option of growing their own fuel has yet to be taken seriously. The ideal crop for 'energy farming' is one with a high energy content and a low processing requirement for conversion to liquid fuel. These crops are in two main groups: (1) high carbohydrate (sugar) crops for ethanol production, and (2) high oil crops. The first group includes various grains (maize, sorghum, wheat) and a range of root or tuber crops (cassava, taro, yams, sweet potatoes). The second group includes sunflowers, soybeans, peanuts, rapeseed (canola), safflower, linseed and cottonseed. Of these, sunflower and rapeseed are high oil producers as can be seen from the comparison of crop oil content and yield in Table 12.1. Presently sunflower constitutes more than 50% of all the vegetable oils produced in Australia.

Table 12.1 Oil content and yield of oilseed crops

Seed	*% Oil**	*% Meal*	*water regime*	*Seed yield t/ha*
Sunflower	40	57	dryland	0.75
			irrigated	1.50
Soybean	18	78	dryland	0.80
			irrigated	2.50
Peanuts	37	60	dryland	1.70**
Cottonseed	18	46	irrigated	1.60***
Rapeseed	39	56	dryland	0.80
			irrigated	1.50
Safflowerseed	34	62	dryland	0.70
			irrigated	1.30
Linseed	37	60	irrigated	1.50

* Assuming solvent extraction
** Includes kernels and shells
*** Assumes 5 bales/ha cotton
Source: Twyford, 1980

12.3 Reducing agricultural energy demand

Besides changing energy sources to home-grown oil or carbohydrate crops, fuel use can be reduced by changing farming methods. Comparisons made in Northern Victoria, for instance, showed that whereas conventional tillage consumed 1,222 MJ/ha of fuel energy to get a grain crop to the sowing stage, minimum tillage consumed only 330 MJ/ha to reach the same stage of production.

The diesel engine is about 33% efficient compared to the petrol engine at 25%. Turbocharging increases diesel efficiency a further 61%. Fuel consumption efficiency can be further increased by correcting the ballast of two-wheel-drive tractors to reduce wheelslip to 8–16% (or 6–12% for 4WD). Of special importance is the setting of tillage implements no deeper than necessary to carry out a cultivation operation (Brown, 1980).

It is sometimes forgotten that reducing total energy consumption on the farm has as much to do with husbandry practices as with direct liquid fuel consumption. An important example of energy saving is the way trees can regulate temperatures in both summer and winter. The benefits of shelterbelts and windbreaks in temperate regions has already been referred to. In subtropical regions substantial metabolisable energy can be conserved by the cooling effect of shade trees. In the case of temperate regions, it has been shown in

Victoria that shelterbelts can increase animal production by 14–16%; in the subtropical case, trials in Queensland showed that lamb survival rates were 10–18% greater in paddocks where shade trees had been provided for lambing ewes.

The ameliorating effects of trees, shrubs and a dense grass cover on the agricultural ecosystem have been well demonstrated under a wide range of conditions. They affect the efficiency of the ecosystem's energy flow, which is in essence a triangle of production, consumption and decomposition. The production processes commence with photosynthesis by green leaves and may be followed by conversion to energy in grain or fruit and vegetables, or the leaf energy may be converted to animal products by grazing herbivores. Since the amount of leaf area available to trap the sun's energy is the key to higher production, energy accumulation is maximised by ensuring the greatest leaf area for the longest period each season. Conversely, if plant cover is reduced — apart from the erosive effects of low plant cover (grain or grazing) — the whole energy cycle of the farm slows down.

12.4 Sustaining energy flows

The use of the sun's energy by rural production systems is regulated primarily by the availability of soil moisture. Thus in the arid grazinglands, sunlight is not a limiting factor, but rainfall is, whereas in humid temperate regions, such as Tasmania, soil moisture is usually sufficient for growth but the capture of primary solar energy depends on the number of hours of sunshine during which photosynthesis can take place. In all cases, sustained maximum production is determined by managing the land in such a way that the total leaf area (plant density) can use the sunlight/soil moisture conditions to the full. Sustainable farming systems can thus be defined as systems which maintain a high level of primary energy capture (assimilation) and conversion. In simple terms, agricultural production combines solar energy with minerals (plant nutrients) and water (soil moisture) to make saleable products. Any management which reduces either the energy capture or the conversion processes reduce productivity, irrespective of the levels of other inputs into the system.

The energy flow of any system is a good measure of its productivity and efficiency. Energy is related to dry matter production through solar radiation and photosynthesis, and its flow through the system is integral to the cycling of nutrients through soil, plants and animals. Energy flow is usually expressed in kcal/m^2/yr, or energy units per unit area over time. Unlike nutrient cycling, which recirculates minerals through the ecosystem, energy flow is unidirectional — that is, it is accumulated and transformed from the sun's energy through various processes to a final breakdown in which carbon dioxide is given off when the organic compounds are oxidised.

In agriculture, the study of energy flow gives an understanding of the rate and extent of energy movement, which in turn explains how the nutrient cycle in the soil/plant/animal complex is functioning. As energy flow accelerates, so nutrient cycling and productivity increase. The rate of the flow is determined by the balance between sunlight, plant material, moisture and nutrients. A very large proportion of the energy present in a cropping system is taken up in the process of respiration — the reverse of photosynthesis (carbon assimilation). Respiration converts energy products such as carbohydrates (sugars) into carbon dioxide. Energy not respired is held in the plant or animal body until it is reduced to its composite elements by decomposers, usually bacteria and fungi.

How do these basic energy conversion processes contribute to sustainable systems? In the first place the primary capture of energy can occur only when large areas of leaf material can trap solar energy. Many crops can produce 21–31 $g/m^2/day$ (Black, 1963) given suitable levels of sunlight and moisture, but most studies make the error of disregarding below-ground daily production in their calculations. Williams (1964) was one of the first scientists to try comparing the energy flows in Australian agricultural systems. Because of the paucity of local available data, the quantities recorded should be viewed as approximate only. Table 12.2 compares *Danthonia* native pastures with subterranean clover pastures under conditions typical for southeastern Australia:

Table 12.2 Energy relationships for native and sown pasture ecosystems kcal/m²/yr

Total solar radiation 1,672,000	*Native pasture*	*Sown pasture*
Energy accumulation		
Net pasture production		
tops	1,880	5,600
roots	500	1,400
Total	2,380	7,000
Energy expenditure		
food	300*	1,880**
excreta	130	790
respiration	180	1,090
	610	3,760
wool	5	40
	615	3,800
Energy available for other primary consumers and decomposers		
tops	1,265	1,800
roots	500	1,400
excreta	130	790
	1,895	3,990
	(80% of net production)	(57% of net production)

* One sheep per acre at maintenance for 12 months
** Four sheep per acre at maintenance for 5 months, gain for 4 months, and loss for 3 months.
Source: Williams, 1964

Table 12.2 indicates the substantial initial differences in energy accumulation between the two pastures (2,380 to 7,000). From this point on the flows maintain these differences all the way through to decomposition.

Within the agricultural systems, energy and nutrients flow in parallel, at least up to a point. The nitrogen cycle is much more complex than the phosphorus cycle and it is much more prone to losses through nitrogen leaching (most N is highly water soluble) and volatilisation (loss as gas, usually ammonia). Pasture systems may contain 1,500 kg N/ha divided between the roots, leaves and litter in an approximate ratio of 11:2:1. This system could cycle 100 kg/ha annually and accumulate a similar amount through symbiotic nitrogen fixation by the legumes. This annual cycle is equivalent to about 7% of the total nitrogen in the system. In the case of phosphorus, the application of 220 kg/ha annually may circulate about 20 kg/ha through the system. Since the export of phosphorus in animal products can only be compensated for artificially, applying phosphate is necessary for the production system. In sustainable systems the role of soil fauna in decomposing organic matter is crucial to both energy flow and nutrient cycling.

Chapter 13

WATCHDOGS AND WHISTLEBLOWERS: MONITORING SUSTAINABILITY

In this chapter we examine the measurement of sustainability of rural production systems over the long term. We consider what constitutes good environmental indicators and how these may be identified for specific use in agriculture. We give special consideration to the problem of chemical contamination and the public perception of the danger of polluted foodstuffs. The chapter concludes by assessing the health of the soil.

13.1 The environmental impact of land use projects

A characteristic of Australian agricultural projects so far is how little they have been subjected to scrutiny by impact studies. Since most environmental assessment has been limited to mining, industrial and residential developments, the environmental impact of many large agricultural developments, including extensive clearing operations, has not been assessed. Under the circumstances, it is useful to examine the assessment stages recommended by the United Nations Environment Program (1988) for any EIA irrespective of size and location:

1 Screening — the initial evaluation of whether the project warrants a detailed assessment.
2 Scoping — the process whereby all the relevant issues requiring inclusion in the EIA are considered and listed, including geographical boundaries, attributes and criteria to be evaluated and timelines for the study and for impact monitoring.
3 Prediction — the estimation of expected direct and indirect effects of the development project on the area's environmental, economic and social indicators. The level of precision and detail should be proportional to the significance of the predicted impacts.

4 Evaluation — the judgements on the significance of the expected impacts as measured against current standards or policy or community values.
5 Mitigation — the consideration of methods of reducing or avoiding impacts, plus the possibilities of compensating for negative impacts.
6 Communication — the interchange of information, values and opinions between the project proposers, the approval authorities and the affected public. This may include documenting the expected impacts of the project, formal objection procedures, opinion surveys and public meetings.
7 Monitoring — checking unplanned impacts during and after construction plus the continual monitoring of the management plan, and how it complies with the specified licence conditions.

As issues such as water quality, habitat retention and integrated catchment management gain more political currency, the requirement for more rural EIS is likely to gain wider acceptance. In recent years the EIS objectives have become part of the normal considerations for planning exercises. These objectives (ANZECC, 1991) include ensuring that decisions use timely and sound environmental advice; ensuring public participation in decisions affecting the community; encouraging design and operation to minimise adverse environmental impacts; monitoring environmental effects, and promoting awareness and values appropriate to sustainable societies.

In the context of agriculture, most of the environmental monitoring requirements are diffuse or small apart from the major production projects on family farms or on company properties and thus not amenable to formal EIS evaluation and public participation. Agricultural impacts require a different system of monitoring: indicators of the effects of farming operations have to be continually checked and measured.

Perhaps the greatest deficiency in present Australian resource allocation policy-making lies in the lack of integrity of Environmental Impact Statements produced under current EIS legislation. The present legal framework stems from the Environment Protection (Impact Assessment) Act (1974) which required developers to identify and evaluate the effects of their operations. State EIS legislation has since followed but the problems of biased statements and lack of monitoring remain. The anti-environmental outcomes of state/developer collusion persist as obstacles to the proper functioning of the community participation process. Also, the current EIS legislation usually relies on the unlikely assumption that the state is impartial and does not affect the conflicting values espoused by the objecting parties. The Tasmanian Wesley Vale pulp mill dispute probably did more to emphasise the need for more rigorous EIS preparation than any other stimulants to the debate on appropriate mechanisms for resource planning and allocation in Australia. In that case study the state government failed to recognise that the community did not accept development at any cost. Today state and federal government policies are starting to reflect more environmentally appropriate community goals such as improved quality of life.

Environmental impacts caused by land use will inevitably become more politically significant as competition for land increases and as community values move from the development-driven approach to a more environmentally friendly approach that values quality of life more than development at any cost. As a prime example of 'policy formulation on the run' the Wesley Vale saga holds important lessons for governments who serve communities who are demanding more input into environmental decision-making. The Tasmanian government of the late 1980s would have done well to take note of what was happening on the other side of Bass Strait where the Victorian government had set up procedures including predetermined principles, guidelines and standards for all development projects before tenders were called. Had such procedures been followed it would not have been necessary for the Foreign Investment Review Board (FIRB) to withhold the pulp mill approval on the grounds of Tasmania's inadequate guidelines.

There is little doubt that increasing competition for land by alternative uses will provoke more intense conflict, notably in the coastal zone of Australia. This will require considerably better mechanisms for conflict resolution than the present largely ad hoc arrangements. Greater pressure will be exerted on the agricultural industry to justify its claim to landscapes that increase their economic and social values as the demand for community amenities expands. Consequently the farming lobby will be able to demonstrate that it is using the land sustainably, and able to claim that agricultural production on the land concerned is at least as high a socio-economic priority as that of its competitors.

There are five categories of conflict resolution (Westman, 1987) for land resource allocation and monitoring:

- Conflict anticipation in which a third party identifies the potential disputes before firm positions are taken up by opposing interest groups. Scoping and screening procedures can be applied in such situations.
- Joint problem-solving, in which a series of informal meetings clarifies and resolves problems met in the early stages of potential conflict. This approach uses advisory committees and dynamic impact assessment with structured citizens' workshops.
- Mediation, in which authorised representatives of competing parties formally negotiate, via a mediator, to reach consensus and agreement.
- Policy dialogues, in which conflicting policy-making agencies meet and agree on policies for implementation by both agencies.
- Binding arbitration, in which an arbitrator is employed to consider and hand down a settlement, the outcome of which all parties have agreed to abide by.

As the Tasmanian Wesley Vale case demonstrated, without clear and comprehensive guidelines and standards for development and regional planning, conflict resolution is time-consuming and generally unsatisfactory to all parties. While it is generally agreed that plans for sustainable development must encompass ecological, economic and social concerns, the intangible values which constitute the 'aesthetic criteria' of the EIS will continue to pose complex problems for decision-makers. In what is called 'landscape evaluation', the public's opinion of scenic values (vistas, waterfalls, mountains) is sometimes gauged by showing photographic montages comparing 'before and after' visual changes.

Table 13.1 Types of assessments and degree of public participation in environmental impact assessment, by jurisdiction

Jurisdiction	*Type of assessment*	*Public participation**		
		Stage 1	*Stage 2*	*Stage 3*
Commonwealth	Public environment report	Fair	Fair/Good	Fair
	Environmental impact statement	Fair	Good	Fair
New South Wales	Environmental impact statement and statement of environmental effects	Fair	Good	Fair
Victoria	Environmental effects statement	Good	Good	Good
Queensland	Environmental impact statement	Nil	Minimal	Minimal
South Australia	Environmental impact statement	Nil	Fair	Fair
Tasmania	Level 1	Nil	Minimal	Minimal
	Level 2	Fair/good	Good	Good
	Level 3	Very good	Very good	Good
Western Australia	Consultative environmental review	Fair	Good	Good
	Public environmental review	Fair/good	Good	Good
	Environmental review and management program	Very good	Good	Good
Northern Territory	Preliminary environmental report	Nil	Minimal	Minimal
	Environmental impact statement	Nil	Fair	Minimal

* Public participation ratings are influenced by such factors as provisions for mediation, public hearings or meetings and frequency of use of these participative mechanisms. Stage 1 involves access to raise all relevant environmental issues; Stage 2 involves access to influence impact study; Stage 3 involves access to ensure that major recommendations are followed.
Source: Environment Institute of Australia, 1990

The shortcomings of the impact assessment process as presently applied in Australia have already been mentioned. There are, however, important differences emerging in the way in which EIA is made in various states. Each state has its own Act under which EIA is administered by a nominated department. In 1990, when a comparison between states was made (Environment Institute of Australia, 1990), Victoria was assessed to have the 'best' procedures in place while Queensland had made least progress (see Table 13.1).

There are few elements of the ESD process more important than the manner in which EIA is carried out. In the past rural producers have largely been immune to EIA, but as other competing land uses gain political acceptance over cropping and grazing the EIA process becomes even more crucial to sustainable agriculture and land use. Since the early 1970s (Talbot, 1976), when EIA entered Australian land use decision-making, specialists have been attempting to improve and refine the assessment process.

Most recently Bates (1992) and James (1993) have evaluated the fairness and rigour of Australian procedures. These analysts identify nine elements of current EIS mechanisms which require significant improvement if future assessments are to achieve their objectives:

1 *Lack of integrity of the EIA process*: Lack of integrity in EIA reports is probably the most persistent of criticisms, leading to allegedly biased and incomplete outcomes.
2 *Narrow project focus and reactive approach*: The absence of links between the assessed project and broader regional development issues makes balanced evaluation almost impossible in many cases. When such isolated project consideration occurs in a reactive mode and (belatedly) examines only one option in land use, the process becomes dubious and does nothing to engender confidence in the outcome or acceptance of the decision.
3 *Deficient integration with management of the strategy*: Apart from the problem of the integrity of the process itself, the absence of (or superficial) links with broader strategic management consideration makes it difficult for planners with perspective to relate the EIS to the holistic approach required in regional planning.
4 *EIA too late in the decision process*: In too many cases, developers first prepare their financial and technical feasibility studies, and only as an 'afterthought' carry out environmental audits, so that the post-planning EIA reports are often little more than a minimal attempt to meet legal planning application requirements.
5 *Insufficient community input*: Public participation is provided for in legislation for all state procedures, but all too often the opportunities for citizens to contribute are minimal and/or their comment is not taken seriously. This shortcoming invariably leads to hostility and suspicion, and the outcome tends to lack credibility.
6 *Underestimation of cumulative effects*: Very few of the current state procedures (except perhaps Victoria) give sufficient attention to the future cumulative effects of developments on both the environment and the community (social impacts).
7 *Unacceptable costs and delays*: From the proposers' viewpoint the additional project cost incurred by having to make an EIS has become a contentious issue which is made worse by the unacceptably long delays in project commencement while due process is followed.
8 *Bias in assessing social and ecological impacts*: There is now a widespread public impression of bias favouring the financial

and economic benefits of development proposals. An impression that adverse environmental effects are underestimated is also prevalent.

9 *Insufficient monitoring*: Most EIAs lack provision for the continuous monitoring of environmental impacts. This creates a justifiable perception that the whole assessment exercise is a 'once-off' hurdle for the developer, rather than a continuous process.

The author would add to the above deficiencies the implications of the consultant who is employed by the proposer to undertake the study. When private consultants seek commercial work from development companies, their submission of adverse environmental reports is unlikely to enhance their success with tenders or their chances of being employed again by that company.

Table 13.2 Prime indicators of sustainable agriculture

Agro-ecological region	*Prime indicators*		
	Management level	*Production balance*	*Resource base*
1 Rain-fed crops & animals	Farm management skills *cash flow, equity, planning	Water use efficiency * yield/area/rainfall	Soil health *pH, nutrient balance, biota
2 High rainfall, pastures for animals	Production/area *animal weight/ hectare	Plant growth/cover *% greenness, species/area	Soil bio-indicators *worm, termite etc. numbers and species
3 Low rainfall rangeland for animals	Management capability *planning, debts, asset condition, record-keeping	Animal health and productivity *live weight gain, quality/quantity fleeces, carcasses	Pasture and soil condition *% bare ground, pasture composition
4 Irrigated crops and pastures	Farm and district *debt, equity etc. for farm level; true cost/ benefit (incl. environmental costs) at district level	Water use efficiency *plant use/water applied, watertable trends, crop weight/ water used	Soil health *infiltration rate, % subsoil compaction, biomass activity, chemical residue level
5 Intensive horticulture & viticulture	% IPM adopted in industry *chemical sales, grower records, faunal surveys	Nutrient balance yield and nutrient contents, fertiliser sales, surface water composition	Soil permeability/ watertables *irrigation water use, piezometry, soil infiltration
6 High rainfall tropical systems	Diversity of production *no. of land uses or crops, no. of isolated vegetation patches	Water quality *surface water composition, blooms, pesticides, sediments	Soil productivity *pH and organic matter trends, subsoil compaction, soil structural condition

* *data or measurement required*
Source: Hamblin, 1992

13.2 Good environmental indicators

The need to monitor progress with publicly funded programs such as Landcare, and to demonstrate financial accountability, has attracted special attention to the indicators of sustainability in recent years. Oakley (1991) suggests that useful environmental indicators should meet several essential requirements:

1. Indicators should be *widely accepted and understood* by the users, the public and the monitoring authorities.
2. Indicators should *form a time series* long enough to detect rates of change and trends, but also over periods short enough to allow early corrective action.
3. Indicators should be *adapted to regional variation* so that impacts significant to particular systems and conditions can be detected and followed.
4. Indicators should be selected to allow *simple and easily interpreted* results to be recorded.
5. Indicators should reflect terms and levels that *users can easily relate to* in their daily lives.
6. Indicators should *relate to available data sets* gathered by methods which are understood and accepted as valid, and which can relate environmental changes to existing records wherever possible.
7. Indicators should show both positive and negative impacts and thus be *balanced* from political and socio-economic viewpoints.
8. Indicators should form a *flexible series* that can accommodate changes in the factors being monitored, or changes in their precision or frequency of measurement, to allow for changed community values and new evidence on environmental safety.

Within the broader ambit of ESD a very wide range of environmental indicators has been proposed, including changes in greenhouse gases, the spread of urban areas and trends in health statistics. Narrowing down the sphere of monitoring to rural land use and agriculture, the environmental indicators can be related to the production systems used in each region. For example, the indicators proposed by the Bureau of Rural Resources workshop (1991) in Table 13.2 are grouped into management, production and resources for each of six major agro-ecological regions.

13.3 Specific indicators of agricultural environmental impacts

Several states have attempted to list the criteria appropriate to monitoring the sustainability of agriculture. The most useful catalogue of indicators to date is that of the Victorian Office of the Commissioner for the Environment, established in 1987. The indicators are conveniently arranged in a matrix of indicators by environmental elements — e.g. fauna, flora, soil, water etc. The Victorian key indicators of soil condition have been extracted and presented in Table 13.3 as an example of monitoring criteria for crop and grazingland.

Table 13.3 Proposed soil indicators of environmental impact

Impact	*Key indicators*
Water erosion	
Sheet and rill	Soil movement • loss (t/ha/yr) • area affected (including deposition)
Gully and tunnel	Length of active gullies/100 ha Area of active tunnelling/100 ha
Mass movement and slumping	Area of active mass movement/100 ha
Wind erosion	Soil movement • loss (t/ha/yr) • area affected (including deposition)
Composite measures of erosion	Soil movement • sedimentation of dams (t/ha-catchment)
Soil structure decline	Pore characteristics Stability of surface soils Soil strength characteristics
Induced soil acidity	Changes in status of soil • pH (H_2O) • pH ($CaCl_2$)
Nutrient status	Total soil nitrogen Nutrients before autumn break Plant nutrients in spring Soil organic matter content Protein content in grain
Soil salinity Dryland Irrigated	Soil: EC sodium chloride Salt tolerant plant species (abundance/ distribution/diversity)
Soil Biota	Total soil organic matter Worm species (species diversity/biomass) Terrestrial invertebrates (micro-arthropods) species (abundance, diversity, density) Bacteria and fungi Soil microbial biomass Microbial respiration

Source: Hamblin, 1992

In the same way as soil indicators are catalogued above, the indicators for surface and ground water are grouped into physical and chemical criteria. Because soil and water conditions must be treated as integral elements of the landscape, sustainability indicators must combine soil and water measurements, particularly chemical components and the buildup of toxins affecting soil and water biology. The water indicators proposed for Victoria are given in Table 13.4.

While there is general agreement that some form of sustainability indicators are required to monitor Australian agriculture, Oades and Walters (1994) suggest that decisions must first be made on the 'scale and hierarchy' in which the monitoring is done, as shown in Table 13.5.

Table 13.4 Proposed water indicators of environmental impact

Impact	*Key Indicators*
Surface waters	
Physical	Flow characteristics
	Temperature
	Suspended sediment
	Turbidity
Chemical	Total phosphorus
	Total nitrogen
	pH
	Salinity (EC)
	Agricultural chemicals in water
	Agricultural chemical residues in sediment
Biological	Fish (native and introduced)
	• species distribution
	• species abundance
	• species diversity
	•. population/age characteristics
	Macro-invertebrates
	• species diversity
	• species abundance
	• population/characteristics
Composite	• volume of water/use type
	• volume of diverted/Basin
Groundwater	
Physical	Volume extracted/rate of recharge
	Groundwater level
Chemical	Nutrients
	Chemical residues

Source: Hamblin, 1992

Table 13.5 Scale and hierarchy in sustainable agriculture

Scale	*Community*	*Indicators*[1]
International	International bodies	State of the economy
	Governments	State of the environment
		• radiative gases
National	Government bodies	• deforestation
		• desertification, etc
Regional[2]	Policymakers	Rural economy
	Consultants	
	Industry-agricultural and manufacturing	Land and water quality
Catchment	Local government	Biodiversity
Farm property	Community groups	Net farm income
	e.g. Landcare	Soil quality
Paddock	Property owners	Modern farming practices

1 Each of these broad indicators can be divided into a large number of individual measurable parameters, attributes, characteristics or specific indicators which at the small scale will have different relative weightings from one area to another.

2 Regions may be defined ecologically or politically, but from a production point of view the ecological basis appears preferable.

Source: Oades and Walters, 1994

The Standing Committee on Agriculture and Resource Management proposed four types of indicators: three onsite indicators of environmental, financial and social measures, and one offsite indicator for environmental effects. Oades and Walters list 24 measurable soil quality indicators for land managers to use in cropping systems. From these indicators they select those that can be used in the paddock without scientific equipment. Oades and Walters realise that, as for as any new approach, their proposals will not gain acceptance until farmers acknowledge the need to monitor 'paddock health'. They propose gaining farmers' interest by using profit

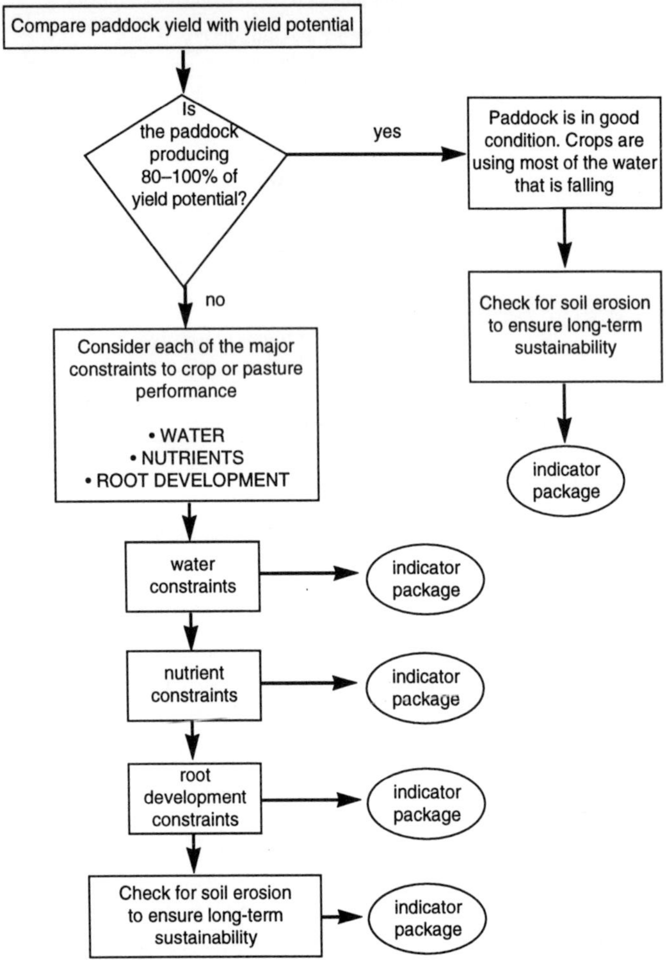

Figure 13.1 Flow diagram showing sustainability criteria and indicator packages for monitoring cropping systems

Oades and Walters, 1994

margins (as reflected in potential yield) as the basic measure, then determining how reduction from the potential yield relates to soil moisture and sunlight factors. They suggest that many paddocks produce less than 50% of their potential yield because of management limitations and propose the attached flow chart to indicate the sequence of analysis by which packages or groups of indicators can check where yield limitations occur, and whether soil loss is acceptable (see Figure 13.1).

Making indicator techniques user-friendly is likely to be the most important attribute of any proposed method. Widespread acceptance of manager-oriented indicators (or indicator packages) may require at least some form of information transfer and self-education. One of the most widely promoted methods of estimating sustainability trends is the 'Right Rotations' program in South Australia (1993). In essence this program allocates points to practices in accordance with the assumed benefit of each good practice. Instead of measuring the final effects of agronomic methods on soil and crop attributes, Right Rotations gives indirect credit for each recommended conservation practice and makes a simple summary of the overall health of the cropping system in a scoresheet which can be used to compare paddocks, properties or even catchments (see Table 13.6).

Table 13.6 Point conversion chart for management practices

	Management practices	*Point score*
Rotation	Clean vigorous legume pasture (>95% legume)	5
	Legume dominant pasture (>75% legume)	4
	Grain legume	3
	Legume/grass pasture (about 50% legume)	
	Natural or weedy pasture (<25% legume)	2
	Cereal/legume mixed crop	
	Cereal, oilseed, etc.	1
	Long fallow started before August	
Tillage	Direct drill, narrow points	0.5
	High-speed harrow	0.5
	Blade plough, rod weeder	0.5
	Direct drill, full cultivation	1
	Each additional cultivation	1
	Offset discs	1.5
Residue retention	Total retention: (over 90% retained) e.g. non-grazed pasture, green manure	4
	Most retained: (65-90% retained) e.g. lightly grazed pasture or cereal	3
	Part removed: (35-65% retained) e.g. grazed pasture or cereal stubble, lightly grazed grain legume	2
	Most removed: (10-35% retained) e.g. baling, heavily grazed pasture, partial burn, grazed grain legume	1
	All removed: (less than 10% retained) e.g. complete burn	0

Source: Herrmann, 1993

Once each paddock is scored, the estimated level of sustainability is determined by where the total score falls in the 10-1 index which is grouped into five major levels of sustainability, three of which are unsustainable (see Table 13.7).

Table 13.7 Tabulated continuum of sustainability scores

Index	*Indicator colour*	*Sustainability*
10 9	Dark green	Highly sustainable. Long-term improvement in fertility and structure, with very low risk of degradation or erosion. Little or no additional nitrogen required. High earthworm activity.
8 7	Light green	Moderately sustainable (15–20 years). Soil fertility and structure, productivity and grain protein is being maintained or improved. Some nitrogen fertiliser may be required. Risk of erosion at critical times in the rotation. High earthworm activity.
6 5	Yellow	Not sustainable. Decline in soil fertility and structure within 5–10 years. Loss of productivity and grain protein within 10–15 years. Nitrogen fertiliser required on most cereal crops. Soil is vulnerable to degradation and erosion. Moderate earthworm activity.
4 3	Orange	Not sustainable. Decline in fertility and structure within 5 years. Productivity and grain protein will decline within 10 years. Nitrogen fertiliser required. Soil is very vulnerable to degradation and erosion. Low earthworm activity.
2 1	Red	Not sustainable. Severe decline in fertility, structure and productivity within a few years. High levels of nitrogen fertiliser are required, and soil is very vulnerable to degradation and erosion. Minimal earthworm activity.

Note: Local knowledge, experience and conditions must also be considered when interpreting this index. High levels of nitrogen fertiliser to maintain production will increase the rate of acidification of certain soils.
Source: Herrmann, 1993

13.4 Chemical contamination — fear of the unknown

The soil and water indicators listed in Table 13.7 were selected to reflect the composite biological health of the soil/water system. In the case of chemical contamination, the aim of monitoring is to combine a series of indicators which reflect how accumulated toxins may affect the health of fauna, livestock and humans. The contamination of both the ecosystem and the agricultural produce it yields is monitored. The key indicators and the secondary indicators of contamination proposed for Victoria are listed in Table 13.8.

Chemical contamination warrants special monitoring since the largely unknown effects of pesticide sprays, in particular on human health, have attracted increasing public concern. For example, the original effects of DDT were so widely publicised that it was banned for use in food crops during the 1970s. The fear of chemical health hazards is not limited to insecticides, however, and in recent years controversy over the effects of herbicides such as Agent Orange in Vietnam on the health of war veterans has made the public question the safety of crop and garden weedkillers.

The term 'chemical' has come to be misused in health and organic publications. For instance, descriptions such as 'chemical-free' are used to indicate an absence of synthetic chemicals, not of naturally occurring compounds in plants and animals. To avoid unfounded emotional claims about agricultural chemicals, particularly the synthetic *biocides*, a public education program is necessary to explain the toxicity, distribution, disposal and reduction of risks for each chemical.

The concern about synthetic chemicals is not only that they may be toxic but that nature may be unable to break them down; for

Table 13.8 Key and secondary indicators of chemical contamination

Impact	*Key indicators*	*Secondary indicators*
Contamination of native fauna	Residues in flesh Morbidity and mortality characteristics	Change in species population characteristics, abundance and diversity
Additional for birds	Eggshell thinning	
Additional for benthic macroinvertebrates	Changes in population diversity, abundance and distribution	
Contamination of soil	Residues in soil • fertiliser residues • persistent biocides	Areas of known use of persistent biocides Areas of current use Residues in surface and ground water
Contamination of produce		
Crops	Chemical residues in crops	Area, volume, distribution of pesticide use
Livestock	Chemical residues in livestock	Area, volume, distribution of veterinary chemical use
Human health (farm workers and rural communities)	Correlation between degree of exposure to agricultural chemicals and onset of disease symptoms Human breast milk surveys for bioaccumulative agricultural chemicals Incidence of acute poisonings with pesticides Incidence of chronic poisonings with pesticides	Drinking water quality (i.e. for chemical residues) Air quality during spraying (i.e. for chemical residues)
Composite measures		Volume/date/area/type of chemical sales Area sprayed — aerial.

Source: Hamblin, 1992

example, they may be *persistent* (like DDT) and become a permanent component of our ecosystems. Currently about 70,000 different chemicals are used in agriculture, in industry and in household products worldwide, according to the British Medical Association (1990). Under the circumstances it is not surprising that public anxiety, the so-called *chemophobia*, is becoming widespread. While agriculture is targeted for its pesticide use, many common consumer goods are considered health risks, often linked to cancer or allergies. Table 13.9 gives a compilation by Silvey, Simpson and Silvey (1993) of Crone's (1990) data on the adverse effects of common organic chemicals.

In agriculture the sustainability of productive ecosystems was threatened by the persistent and chronically toxic biocides termed *chlorinated hydrocarbons* such as DDT, BHC, Aldrin and Dieldrin. These stable compounds resist degradation and are now banned in Australia and most developed countries. Pesticides have generally helped to keep down pest damage to 5–30% of the potential production of most crops affected by pests, but there is growing evidence that pesticides seriously threaten the health of farm workers. The health hazard to consumers of pesticide-treated crops has been less conclusively proven. Pesticides also affect the honey bees and wild bees responsible for pollinating many fruits, vegetables and forage crops. The US Environmental Protection Authority currently

Table 13.9 Perceived adverse effects of some synthetic chemicals

Class	*Description*	*Examples*	*Adverse effects*
1 Structural Chemicals			
Bulk plastics	Used for implements, furniture, plumbing	Perspex, PVC, polystyrene, plastic containers, polythene bags	Aesthetic (visual), possibly economic (damage to useful plants and animals)
Fibres	Used in textiles, carpets, ropes	Rayon, nylon, terylene, fishing line	Aesthetic, economic
2 Pesticides	Control pests such as insects, weeds, fungus and snails	Parathion, carbamates, 2,4,5-T, metaldehyde, used containers	Health (deleterious to humans and non-target organisms)
3 Process Chemicals			
Industrial	Used to modify or facilitate a process without forming a significant portion of the end product	Wetting agents, mould release agents, PCB, PBB, sterilants, enzymes, dioxin, additives	Health, economic
Domestic	Variety of uses, with some substituting for natural products	Detergents, polishes, disinfectants, cosmetics	Health, economic
4 Drugs	Chemicals for treating disease or which affect body functions	Antibiotics, antidepressants, aspirin, salbutamol	Health (environmental amounts negligible)

Source: Crone, 1990

restricts the use of over 100 major pesticides specifically because of their effect on about 250 endangered species in 900 counties across North America. They are also seeking more appropriate control measures for 440 insect and mite species which have become resistant to common pesticides since the 1940s when DDT and Parathion were introduced. Clearly the use of more frequent and higher rates of current insecticides is not the long-term answer.

The human health issue over farm chemicals has, predictably, received more attention than issues such as domestic stock or wildlife health, including the economically important group of insect-eating birds. Apart from cancer, agricultural chemicals have been linked to birth defects, behavioural abnormalities, immune system dysfunction and allergic reactions. The greatest hazard is from the strongly persistent organochlorine pesticides. The lack of accurate exposure data (dose, frequency and time of exposure) and incomplete or even conflicting data on health effects have been major obstacles to developing standards and giving pesticide users and sprayed food consumers confidence in the safety of the chemicals used in modern agriculture (NRC, 1989). For instance, fungicides pose a very difficult problem because few alternatives are available. Fungicides pose a greater cancer risk than insecticides and herbicides combined, but are still used on a vast scale for the production and storage of foods.

Any review of hazardous chemicals in agriculture should include the health risks of organic fertilisers such as manures and sewage sludges. Hazards include higher N (nitrate-form) in foods and water. The derivates of nitrate may be poisonous or carcinogenic or both. Some sewage sludges contain high levels of heavy metals such as cadmium, chromium and lead, which can accumulate in the soil and in living tissue. Human pathogens can also be spread in sewage and compost. Australian data on these health hazards is difficult to obtain, but the US figure is 33 million illnesses and 9,000 deaths caused annually by food-borne illnesses, of which many can be attributed to pathogenic bacteria of animal origin such as salmonella. Little data is available on the effects of antibiotics and growth promotants on human health — perhaps a case for the application of the 'precautionary principle' referred to in section 3.4.

13.5 Chemical resistance and soil health assessment

Sustainable agriculture will eventually have to be based on lower inputs of chemicals. One of the more costly aspects of modern agriculture is the rate at which pests, diseases and weeds tend to develop resistance to the ever-increasing range of chemicals synthesised to combat them. In Australia this is most noticeable in pests such as cattle ticks and several insect pests of grain crops. Scientists have been surprised at how fast these organisms can adapt to chemicals that initially eliminated them. Whether by mutation, natural selection or a flexible metabolism, these species develop an almost complete resistance to many chemicals (applied as sprays) within several seasons, and so it is now common practice to advise against using any chemical for more than one or two seasons. This prevents the organisms from developing a natural selection process against a specific chemical.

Several methods of monitoring soil conditions have been put forward — for example, the Agricultural Monitoring Sites network which is based on a grid system representing all major soil types and production systems. A world soil health index has also been proposed. Herberen (1992) tells us that soil quality is much like the weather: 'people talk about it, few understand it and hardly anyone does anything about it'. The Rodale Institute in Pennsylvania believes that inherent knowledge of the soil is not enough; we need some form of report card to indicate trends in gains and losses in soil quality. A soil index could guide agricultural research, focus on priority regions and problems, and help forestall environmental disasters, famines and chemical hazards. It could determine where soil quality is improving so that the lessons learnt there can be applied to needy areas. Could the Australian Working Party's (ESD, 1991) soil indicators be used in such an index? They offer a comprehensive series of criteria for soil health (quality, condition) which could be much simplified and standardised for international use and comparison.

Chapter 14

FARMING SYSTEMS RESEARCH

This chapter examines the need for research that will enhance the understanding and development of cropping systems that can be maintained permanently. We examine how operations research and modern computer modelling contribute to sustainable agriculture. Now that the goals of research programs have moved from high production to maintaining the soil resource, renewing soil with soil forming processes has become possible.

14.1 Components and priorities in sustainability research

Earlier agricultural research in Australia concentrated on higher production from each enterprise or commodity such as wheat, wool, beef, etc. Over the years researchers came to recognise not only the links between different components of the total farm production system, but also their inability to solve degradation problems individually. Once it became clear that the interactions between components such as soil, crops, animals and water were as important as each component's processes, 'systems thinking' became accepted by researchers as the only way of approaching agricultural research.

This approach grew out of the General Systems Theory used to increase efficiency in many industries in the 1960s, but its application to agriculture led to much greater complexities and uncertainties in this applied field. Agricultural systems are not only practised under unpredictable weather conditions and highly variable cost/price regimes, but include biological, chemical, physical and financial subsystems. In simple terms, systems researchers were dealing with people's reactions to their environment, and the responses of that environment to management. So they had to know not only what makes up the agro-ecosystem and how it functions, but how this system interacts and how it responds to the socio-economic system which intervenes with management inputs.

As a guide for systems researchers in agriculture, Squires (1991) suggests that because of the complicated nature and varied interactions of the soil-water-crop-animal complex, four assumptions can be made to simplify the research approach:

1 It is not necessary to know everything about a system to be able to produce a useful analysis of that system.
2 Understanding the important properties and reactions of a system requires a good comprehension of only certain basic relationships within that system.
3 The productivity and efficiency of a system can be greatly improved by a few fundamental changes in management.
4 Understanding which changes are likely to be fundamental to the system requires formulating and correctly answering certain key questions.

The era of the great advances in agricultural research may have passed forever. The early increases in production gained by new fertilisers (superphosphate), new pasture plants (subterranean clover), new animals (humped cattle) or new machinery (Ridley's stripper) are unlikely to be matched by future research. The specialist (reductionist) approach of the past allowed researchers to concentrate on a single limiting factor in the production system. By improving one element, spectacular increases in yield could be gained.

McCown (1991) explains how earlier fragmented research in specific disciplines has given way to teams of researchers. Problem-solving in complex production situations was not as attractive to agricultural researchers as the narrower research projects in which they could show their originality and gain the respect of their peers. The question of what was worth researching on a cost/benefit basis was seldom asked when researchers set their own priorities. Today two major factors have influenced research priorities: (1) emerging environmental concerns, and (2) greater scrutiny by producer organisations, which provide the bulk of research funds. As a result, there has been a shift away from research aimed primarily at yield increases, and a new emphasis on what is termed Farming Systems Research (FSR). Such systems research gives priority to the wellbeing of the farming family, the farm and its resources, and the communal environment affected by farming practices.

An analysis of the so-called 'production ecology' of the farming system may start with three of its attributes: *stability*, *productivity* and *sustainability*. Scientists from the physical, biological and social sciences may be required to team up if a holistic approach to the projects is envisaged.

The quest for reliable sustainability indicators reviewed in Chapter 13 has led Hamblin and Goss (1993) to identify and select indicators that can be reliably estimated and objectively compared —

that is, indicators useful for sustainability research. Their matrix (Table 14.1) indicates the relations between their four prime indicators and the major attributes of sustainable systems.

Table 14.1 Indicators and attributes of sustainable agriculture

Key indicators	*Major attributes*
1 Long-term net real farm income	Net farm income, productivity, terms of trade and numbers of farms over time
2 Land-water quality to sustain production	Water use efficiency of crops and stock, nutrient balance of farm, area of native vegetation and degree of fragmentation of vegetation
3 Managerial skills	Farmer education level, skills index, Landcare attitude index, and farm planning capacity
4 Offsite environmental impacts	Food chemical contamination level, river outfall turbidity, dust storm frequency, and length of contact zone with non-agricultural areas

Source: Hamblin and Goss, 1993

The Hamblin and Goss indicators are yet to be fully tested under a range of agro-ecological conditions. From a research viewpoint they may be more reliable criteria than other, purely physical or biological indicators. The crucial question is whether researchers have identified the prime attributes of sustainable agriculture applicable across Australia's range of conditions. Hamblin and Goss have chosen 16 attributes which cover income, land quality, management and downstream effects and they maintain that sufficiently reliable comparative measures of these attributes can be made to reflect overall sustainability. Research planners should give these attributes serious consideration.

In the case of sustainability research for the rangelands of the arid zone, a four-tier program is proposed by Morton et al. (1994) mainly to preserve biodiversity — pastoralism has been dealt with by Pickup and Stafford-Smith (1993). The first research program is aimed at understanding the distribution and function of the organisms constituting the richer and poorer landscape areas. The second program researches the extinction process in resource-rich land — natural and managed — under stress. The third program concentrates on the recolonisation process after local extinctions, and the fourth examines how management can reduce extinction and increase colonisation, and includes ecological, economic and social implications. From these four programs, sufficient ecological understanding may be gained to develop sustainable management of the fragile arid rangelands.

14.2 Systems research and operations research

Having accepted the advantages of a systems approach, researchers usually realise that the structure and the management of the system can be better understood if Operations Research (OR) is employed to determine how to improve the efficiency of the production system. It has been suggested that despite OR's major potential contribution to agricultural research, it has received very little acknowledgement in this rural sphere. There are of course close similarities of objectives and methods between OR and some aspects of FSR in which whole systems are represented with various techniques such as mathematical models and other models to demonstrate complex relationships and interactions. For example, McCown (1991) has constructed a model showing how operations research can identify weaknesses in a summer cropping system. Besides the crop production processes the model shows also the climatic influences and economic returns. This kind of operations research can answer practical questions not only about production problems but — more importantly — about sustainability and whether the particular type of farming should be discouraged because of the degradation it causes.

The concept of OR has been used since the 1930s, and modelling has been a research tool since the 1960s. Economists have always appreciated these methods but agricultural scientists have had great difficulty in developing models that produce useful results at a reasonable price. This difficulty persists, and still perplexes researchers and their funding bodies. However, a range of models now available for animal and crop production has led to considerably greater understanding of the systems concerned, even if the quantified data is less than satisfactory from a scientific point of view.

14.3 Modelling sustainability

The previous chapters discussed the complexities of determining and monitoring the sustainability of a system. Such complexity is too great for the human brain to comprehend, but the computing and memory power of modern computers allows this complexity to be fully analysed for the first time in the history of agricultural research. One of the best examples is the Model of an Integrated Dryland Agricultural System (MIDAS), developed by Morrison, Bathgate and Barton (1991). This is not a simulation model, but a mathematical programming model developed by scientists, consultants and farmers to identify the sustainability components of Australian cropping systems and warrants close consideration.

The MIDAS model is based on nine major components, each representing a management decision-making option:

1 *Rotations*: 20 different combinations of cereals/pastures/legumes are compared on seven different land management units. In effect, 80 cropping combinations are compared on each soil type.

2 *Crop yield and response to fertiliser*: the model treats phosphorus as a standard requirement but varies nitrogen from 0–50 kg/ha/yr.
3 *Cropping machinery*: different machinery options are compared between different runs of the model.
4 *Sheep production*: 30 classes of sheep are used to vary flock structure, from which liveweight, lambing and selling times are computed. From these data the most profitable flock structure, sheep numbers and stocking rate are identified.
5 *Fertiliser application*: apart from the crop response component, the cost of optional types and levels of fertiliser application are compared.
6 *Sheep feeding options*: alternative pasture, crop stubble and grain feeding strategies are compared in terms of cost and weight gain. Carryover feed and changes in quality are accounted for.
7 *Product sales*: grain, wool and sheep selling options are compared.
8 *Finance options*: borrowings on various terms, and interest on investment are compared for returns.
9 *Overheads*: the model accounts for fixed costs associated with necessary farm infrastructure and operations, including maintenance and replacement costs.

How does the model account for the sustainability of the system's operations? The MIDAS approach limits the consumption of pasture and crop stubble to a level compatible with soil stability against wind and water erosion. In its original form, it does not measure the other sustainability indicators related to chemical or toxin buildup in the system. However, these can be added as further checks on the modelled system's sustainability. The more sophisticated versions of MIDAS can identify the effects of various management options on land degradation, especially salinisation and wind erosion trends.

There are welcome indications that Australian agricultural research is moving away from the earlier emphasis on yield and toward a systems approach in which the environmental impacts and thus the permanence of comparable systems are being understood and counteracted.

14.4 Other uses for modelling

Despite a rather widespread disillusionment with models after the early enthusiasm for them, they have proved useful in a range of spheres that cannot readily be researched by alternative means. Rickert and McKeon (1991) have given a well-balanced overview of the strengths and weaknesses of models for evaluating agricultural production systems. They suggest that once the early problems of insufficient and unreliable data were overcome, models became useful for predicting the best directions for new research, for comparing and assessing research proposals and for evaluating the sustainability of regional farming systems. The model called PERFECT was used to study dryland cropping techniques in Queensland in

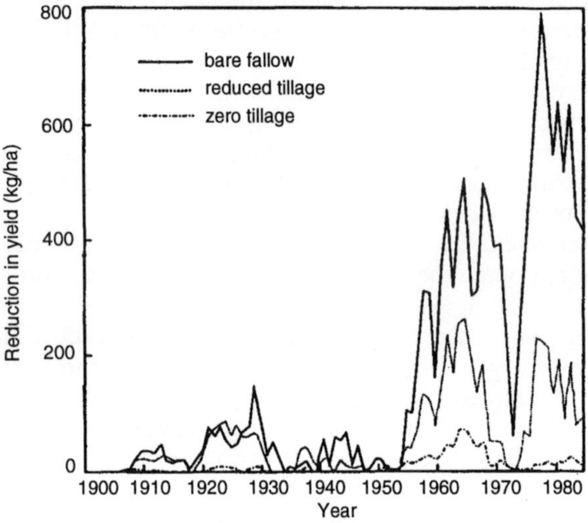

Figure 14.1 Crop yield data relating management systems to predicted soil loss in Central Queensland, as used in the PERFECT model
Rickert and McKeon, 1991

terms of a system's productivity, reliability and sustainability. This kind of computer modelling uses quantitative data and is classed as 'hard' systems analysis, as compared to 'soft' systems analysis which uses qualitative criteria, value judgements, attitudes and perceptions. The mathematical programming used in agricultural models has grown out of the earlier process of linear programming which was devised to identify the optimum combination of factors necessary for a specific goal — for example, highest yield, greatest profit or most biologically sustainable system. Today the Queensland Department of Primary Industries offers landholders a range of management aids in the form of predictive models. These include WHEATMAN which compares cropping strategy options, GRASSMAN which analyses grazing options, and RAINMAN which evaluates management options in a range of weather patterns and soil moisture situations. An example of how the PERFECT model simulates the effect of alternative tillage methods on soil loss and crop yield over a 70-year period is shown in Figure 14.1.

In the sphere of agricultural chemicals and pesticides, models such as the LUPEST program of the New South Wales Department of Agriculture can predict insect problems by using the 'if. ... then' logic which compares a range of predictions and matches each with advice on minimal pesticide applications.

Perhaps the potentially most important function of modelling is its role in education where it can enhance 'systems thinking' more than any other educational technique. Special models have been devised for problem-solving exercises which may be interactive or may present an automatic simulation of a system. Interactive models put the students in the decision-making role so that they can see the outcomes of their management decisions, taken for a specific set of climatic and economic circumstances. This technique is particularly useful in teaching risk management, especially when the model combines financial decision-making with physical elements of the production system, as shown in Figure 14.2.

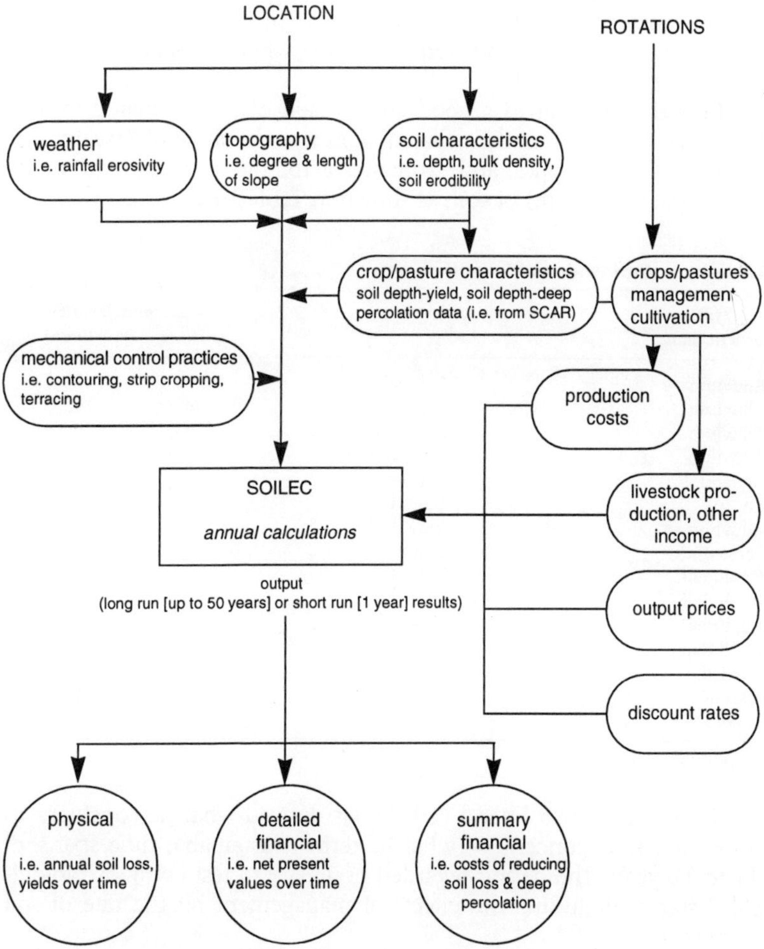

Figure 14.2 Combined physical and economic SOILEC model for Australian crop systems

Elveld, Johnson and Dumsday, 1983

14.5 Renewable soil — does the concept work?

Examples of soil-building and soil renewal have always induced optimism in producers and conservationists alike. The rate of soil formation has been generalised and usually shown to be very slow, at least in human economic terms. Thus while the process of pedogenesis is well understood (Jenny, 1941) the usefulness of the concepts of renewable soil depends almost entirely on its rate under certain conditions of substrate, temperature and moisture. This concept is central to the sustainability debate in agriculture. Grierson (1985) has suggested that Australian soil scientists should group soils as renewable or non-renewable so that the urgency and priorities of conservation farming can be better evaluated, especially where soils are both shallow and non-renewable.

Friend (1992) gives a good comparison of the estimated formation rates of various soils, collated from a wide range of research on several major soil types, and expressed as the rate in years required to form 1 inch (2.45 cm) of soil, as shown in Table 14.2.

Table 14.2 Worldwide soil formation rates

Soil type *Generic name*	*Location*	*Formation rate (years/inch)*
Alluvium	Colorado, US	271
Alluvium	Pakistan	190
Alluvium	St Vincent	44
Alluvium	Wisconsin, US	203
Argiaquoll	Oregon, US	21
Entisol	Hawaii, US	3
Alluvium	Iowa, US	8
Ferosol	Senegal	6
Hapludalf	Iowa, US	102
Hapludalf	Iowa, US	211
Hapludalf	Wisconsin, US	97
Hapludalf	Iowa, US	30
Mollisol	Iowa, US	76
Oxisol/s	Africa	1,905
Spodosol	Europe	53

Source: Friend, 1992

From these data Friend (1992) has derived what he considers an important new concept — what he terms 'sustainable time spans' of 11 to 15 years, the period regarded necessary to test complex biological systems, including the effects of management on the rate of soil formation.

Chapter 15

RURAL WOMEN: NURTURING THE LAND

This chapter moves from the technical aspects of farming to the community value of nurturing the land, and how women's values counterbalance men's domineering and aggressive attitude toward land resources. The chapter deals specifically with the major contribution women have made to the Landcare movement. We explain how women's feminine values can form a basis upon which the whole community can build a mutually beneficial symbiosis between humans and land, as the foundation of land stewardship in Australia.

15.1 Feminising land management values

It is understandable that the image of the tough independent pioneer still persists as the kind of person required to handle the rigours of life on the land. The *frontier mentality* was born out of the inherent survival instinct of those faced with opening up new lands. The need for this view has diminished as agriculture has moved from a physical activity requiring strength and masculinity to a business requiring long-term planning and an appreciation of both the ecological and risk elements of farming. As one Northern Territory manager of a large beef station put it: 'Goodbye macho, welcome femino', when referring to the shift from the domineering male attitude to land, vegetation and animals to a symbiosis or interdependence with the natural resources.

One of the obstacles to communicating the idea of developing agricultural systems in equilibrium with nature is the deep-seated notion that humans have been set on this earth with a divine right to use and have *dominion over* (Genesis 1) the earth and all that is in it. For a fuller critique of this 'God-sent' right to dominate nature, see the writer's challenge to the Rural Ministries of Australasia (Roberts, 1992).

Before considering the important contribution women can make to Australian land management, it is worth analysing the philosophical significance of *values* in natural resource management generally. Nowak (1992), a rural sociologist of Wisconsin, has given a challenging critique of present values. He believes that we in conservation farming waste far too much time preaching values among the true believers and the converted instead of focusing on individuals who are abusing the land. More time must be put into understanding why behavioural change does not follow our sermonising. Nowak does not 'blindly accept the importance or usefulness of present dominant values such as stewardship or sustainability', but he believes that values operate at two distinct levels in land resource management. The first level expresses our objectives — stewardship and sustainability. The second level of values is more useful since it describes how we achieve our objectives. In the Australian context, Nowak would ask us to expand on the values required to achieve Landcare. He wants us to separate 'policy objective values' such as *equity*, *efficiency*, *effectiveness* and *knowledge* itself.

The idea of equity cannot really be interpreted as equality in land management, except in the sense of sharing benefits fairly among individuals or groups. Rather, we should refer to 'equitable access to stewardship and sustainability'. 'Efficiency' and 'effectiveness' refer respectively to output per unit of input, and degree of attainment of objectives. As for 'knowledge' as a value, it must replace the emotional appeals and political manoeuvering by clarifying what sustainable land use actually embraces. Scientific knowledge must be used to define the boundaries of stewardship and sustainability. In promoting desirable values in land management, Nowak says:

> I do not want to be subjected to another layperson sermon on the moral imperative associated with stewardship, or for someone to use the latest 'buzz-word' of sustainability without spending an equal amount of time telling me how to achieve them ... Continued 'boosterism' of values, such as stewardship and sustainability only serves to obscure contradictions in market and resource policies, to rationalise rhetoric rather than promote reason, and to justify mediocrity in organisations with natural resource responsibilities.

Apart from a land ethic, farmers probably also support and operate a family ethic, a market ethic and a community ethic, all of which reflect their moral responsibilities to different social groups.

15.2 Ecofeminist values and their influence on attitudes

The very word 'ecofeminism' elicits negative and defensive reactions from those who have come to associate the prefix 'eco-' with resistance to progress, and 'feminism' with irrational non-acceptance of their male-dominated society. But in the broader context of gender equity, it has been claimed that women represent the greatest single force of political renewal in a civilisation thoroughly out of balance. There is considerable evidence that Australian politics and society are currently in the process of being *feminised*.

Many sceptics doubt whether it is accurate to ascribe the qualities of caring, sharing, cooperating and concern for nature exclusively to females. Nevertheless there are clear indications that women have been making a disproportionately large contribution to environmental consciousness and to corrective action against ecological collapse. It is claimed that feminism in Australia today cannot be interrupted in its direction, but only in its intensity and rate of progress.

The combination of feminism and ecology has been described as the third wave of feminism and it is based on a simple but controversial philosophy: there is a special link between women and nature; men have tried to control both women and nature — therefore, if society is to solve environmental problems and end the oppression of women, it must adopt less dominating and less masculine values and procedures (Westbury, 1991). Ecofeminism offers not only a way of cleaning up our environment, but a new, more holistic way of thinking. Some would call this a more sustainable worldview, as depicted in Figure 15.1.

Today many rural women would not be comfortable to be called ecofeminists but their commitment and their initiatives suggest that this term describes them well. There are of course feminists who concentrate on issues other than environmental quality — for example, social justice, gender equity, female safety, women's health and women's politics. However, in various countries the community environmental movement has had women as its crusaders — the 'hug-the-trees'

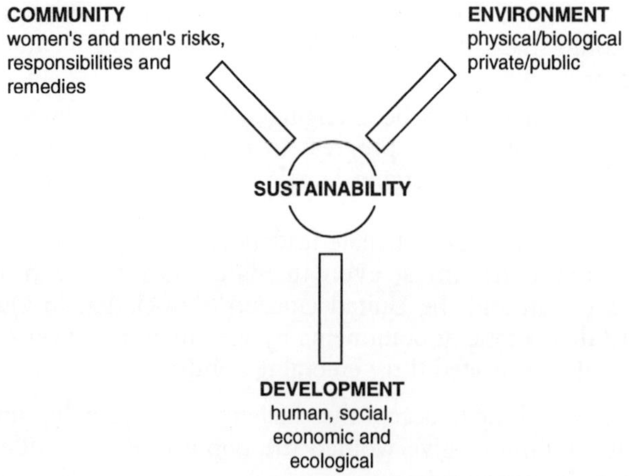

Figure 15.1 A environmental planning framework for community development
Brown and Switzer, 1991

movement (Chipko) in India, tree-planting in Kenya, forest protection in Malaysia. In Australia the Green Party and the Democrats are dominated by women in their strong environmental stance on national politics. Indications are that tomorrow's political leaders will be supported not for their strength or leadership but rather for their intelligence, credibility and creativity. Women who demonstrate unselfish caring attributes and a superior ability to operate information networks and consultation procedures are likely to gain increasing support from voters who recognise the urgency of environmental issues.

While femaleness is not an essential ingredient for environmental commitment, the overwhelming effect of women on environmental group action in communities is the major reason for the emergence of ecologically oriented thinking in today's politics and social policy.

15.3 Rural women and Landcare

The Landcare Movement evolved from an agreement between the National Farmers' Federation and the Australian Conservation Foundation to seek commonality on sustainable land use. This joint venture gave all concerned rural citizens the opportunity to organise their communities for a new and special type of environmental action. In many districts a few men and women had been setting fine individual examples in conservation farming before Landcare was established.

In encouraging rural women to contribute more fully to rural conservation and sustainable production systems, two major changes are required:

1 acceptance by the male-dominated rural society that women warrant greater organisational and leadership opportunities;
2 agreement that if women are to expand their role and activities, men should share the home workload that women have traditionally borne.

While their self-image may be changing significantly, women cannot be expected to make much progress in broader spheres of activity until the image of the emerging female rural leader becomes more widely accepted.

Heartening examples of female leaders are becoming more commonplace in recent times, even in male bastions such as the Cattlemen's Union and the United Graziers' Association in Queensland. These democratic appointments by grassroots producer organisations have demonstrated three emerging trends:

1 Women are willing to accept the challenge of a leadership image.
2 Men are willing to give women the opportunity to guide their largely male organisations.
3 Women are demonstrating their competence and ability to encourage people to join rural organisations and to engender a cooperative spirit in and between such organisations.

Internationally, women have found that it is through Non Government Organisations (NGOs) that they can often make their strongest impact — at least in the initial stages of gaining recognition by governments. In March 1991 the 35th meeting of the International Commission on the Status of Women held in Vienna gave special attention to finding ways and means of integrating women into leadership roles in the world's NGOs. In November that year the Florida meeting of the World Women's Congress for a Healthy Planet provided strong evidence of women's determination to gain more action on global pollution control, clean water supplies and better women's health services.

What has this got to do with Landcare? Very much, if we recognise the urgency of Australia's land degradation problems and the growing evidence that women are poised to lead the conservation and environment debate in this country. With women at the helm of the Ministry for Environment, of the Australian Conservation Foundation, and of national action groups such as Greening Australia, the 'hand that rocked the cradle' is already rocking some traditional strongholds of the economic rationalists. The recent inputs of the Office of the Status of Women (OSW) into the national committee on Ecological Sustainable Development (ESD) must to be taken very seriously by government. In the same way, decision-makers in Australia should take heed of the OSW's recommendations to the Global Assembly of Women and the Environment: 'Partners in Life' (OSW, 1991).

Two outcomes are apparent from Australia's membership of international organisations:

- Networking and swapping ideas with peers in other countries have given Australian women's organisations new information, encouragement and strength to reach their goals back home.
- Comparisons with developed countries overseas have allowed Australians to identify the shortcomings in gender equality and opportunity in a society which has been moulded by the biases of traditional 'mateship' and which excludes half the population from its leadership and direction.

Many writers have been criticised for attributing particular abilities or sensitivities to one gender. The heterogeneity of either gender group may make generalisation dangerous because critics invariably point to the exceptions. However, history and experience tell us that as a group, women do tend to hold certain values more dearly than men do, and tend to be more sensitive about certain issues than men are. When these apparent differences are reported, they are often portrayed by the media as the latest saga in the 'battle of the sexes', as if competition and dominance are the key ingredients.

With land degradation at last recognised by government as Australia's number one environmental problem, the question of environmental sensitivities becomes very important. Are women more concerned about the environment than men are? A 1986 national

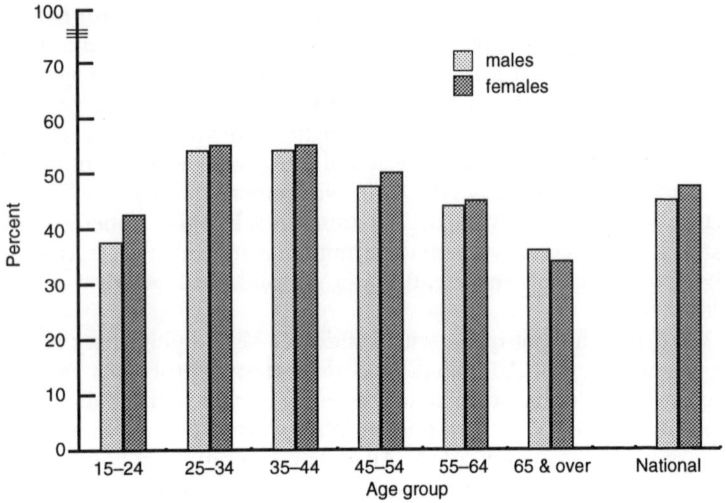

Figure 15.2 Age and gender analysis of environmental concern in Australia
Australian Bureau of Statistics, 1986

survey says yes, at least up to the age of 65 when men seem to get the message and become more concerned than women are, as shown in Figure 15.2.

In the Landcare movement women took an early lead in tree-planting projects, habitat retention programs in schemes such as 'Save the Bush', and in urban landscape improvement activities. In more recent years many women have been elected to chair or manage Landcare Groups in their district. The author has had the privilege of visiting National Soil Conservation Program projects in all states since 1984, and gained from these visits to rural communities the impression that the local women's ability to encourage people to participate and to gain the cooperative support of technical experts has been vital to the success of many Landcare Groups. In many cases women have built a network between interested parties without the competitive or exclusive leadership style which too often has plagued rural organisations in the past. A quick look at the lists of Landcare 'movers and shakers' in various states shows that more and more women are taking a managerial role in Landcare. This movement has the potential to become Australia's most important popular movement ever — it already has over 2,000 Landcare groups nationwide.

15.4 Women and Landcare committee work

While social survey data show that many women's prime activities are somewhat removed from environmental action in rural Australia, the data illustrate the preponderance of women in the activities that maintain the basic health, education and nurturing of our communities. The tendency for women to concentrate their efforts in these fields can be productively used in rural conservation where the development, promotion and acceptance of a land ethic (Roberts, 1991), as the guiding principles in environmentally friendly land use, are crucial to our future. Denzil Mills (1991) has presented the first estimates of the numbers of women nationwide in 'managerial' positions in the Landcare movement (see Table 15.1). While Mills' data are incomplete, they still indicate that although women have a long way to go, they are starting to gain acceptance from federal to local level.

Table 15.1 Women holding positions in Landcare

Queensland: Women holding the position of chairperson or secretary of Landcare groups	
10 out of 107 chairpersons	
35 out of 107 secretaries	
Western Australia: Women on district Landcare committees	
113 out of 650 positions	
Landcare coordinators/facilitators	
Queensland	3 out of 8 positions
New South Wales	2 out of 13 positions
State assessment panels positions	
6 out of 70 positions	
Federal Soil Conservation Advisory Committee	
4 out of 11 positions	

Source: Mills, 1991

The Landcare movement depends on political support for funding its demonstration projects and is gaining support from female politicians at various levels of government. Mills (1991) states that 17% of local government representatives are women. In Queensland this figure increased by 48% between 1988 and 1991. Some 37 of the 460 members of the Australian Rangeland Society are female according to Mills (the society's president).

Women's successes in Landcare should not detract from the excellent initiatives given by men in the same sphere. The good male leaders all demonstrate a leadership style that is acceptable to their peers. For female leaders this task is doubly complex because deeply ingrained role stereotypes have to be broken down in their group. When Fred Astaire was being heralded as the greatest dancer of all time, someone pointed out: 'Just remember that Ginger Rogers did everything he did — but she had to do it backwards, in high heels!'

In some districts women's enthusiasm, drive and ability still tend to be interpreted as 'pushiness', so emerging female leaders often tend to adopt the 'invisible leader style — self-effacing, conciliatory, and letting others take the credit for group achievements. The tendency to interpret attention to detail as 'thorough' in men but 'petty' in women is still widespread. Similarly, the distinction between 'drive' and 'pushy' is regularly made according to gender.

15.5 Sustainable female values

Not surprisingly, survey data on rural women's problems reflect the primacy of social and financial issues in their lives. Within these two spheres of problems, the respondents quoted isolation and low income as their major concerns, but it is the growing stress of meeting family needs in tough economic times that has probably prevented even more able women from taking up the challenge of land stewardship. On the matter of isolation, there is no doubt that group activities such as Landcare, offer attractive social settings for lonely rural women. It would not be surprising if surveys indicated that friendship, group identity and community activity have much to do with women's strong contributions to the Landcare movement. This organisation supplements or even replaces the sports club, church or CWA in the lives of socially active country women, and it probably attracts many men for similar reasons, albeit unconsciously.

Apart from maintaining its food-producing soil, any sustainable society must strike a balance between population and land resources, reduce pollution to levels the ecosystems can cope with, and attain low levels of consumerism and high levels of recycling. Its members must be committed to caring and sharing, living frugally and living gently on the earth. If we are to gain acceptance of these values, the Australian society must move from its consumptive materialism to a more feminine set of values which nurture personal relations and appreciate natural values. In rural districts this will mean changing our agricultural advisory systems to programs based not on technology but on an understanding of rural sociology, personal development and behavioural science. Women's strong influence on our extension services has become vital to their future success.

A decade ago Sara Ebenreck (1983) wrote of the importance of recognising the mutual dependence between humans and 'their' land. She referred to the 'partnership ethic' as a basis for sustainable production and quality of rural life. Integral to such a symbiosis (coexistence) are sensitivities which apparently come more readily to women than to men in general. Gillard, Brough and Duffield (1990) have examined the role of women in rural advisory services and they suggest that women's community involvement often leads to better networking between information sources, cooperators and resource people. They also suggest that maintaining positive and supportive personal relationships is important to women, that women tend to act more intuitively and that lateral and innovative problem-solving is often initiated by women.

Chapter 16

THE POLICIES OF PERMANENCE

This chapter sketches Australia's position in the competitive global market and examines the need to intervene in the free market to protect land resources. We analyse full environmental costing as a basis for resource protection and we use the Western Australian wheatbelt as an example of ecologically suspect agricultural development. As an alternative to traditional policy this evaluates the Bargaining Model as another option for more responsible long-term legal guidelines for rural development and production.

16.1 The international politicisation of sustainable land use

Much of the financial pressure on Australia's agriculture is due to its international market position. For many years about three-quarters of Australia's rural products were exported and in most cases this country was a 'price taker' with little bargaining power. The short-term cost of sustainable agriculture will have to be built into the international trade of rural products if the stability of land resources is to be achieved globally.

Ciriacy-Wantrup (1963) was the first economist to recognise the problems of distinguishing between the technical/economic and the ecological morality aspects of sustainability. By 1980 the International Union for the Conservation of Nature (IUCN) had published its World Conservation Strategy which for the first time maintained that international institutions and national governments are among those responsible for maintaining the ecological health of the planet. This Strategy formed a bridge for negotiation between the conservationists of the developed countries and the leaders of the developing nations who suspected that their potential natural sources of wealth could be withheld from them in the name of global conservation. People on a survival tack have little sympathy for the eco-moralists. O'Riordan (1988) suggests that the international politics of sustainability can now be considered at three levels: basic needs, sustainable use and eco-development. The basic needs form the prime objective, sustainable use offers the means of reaching the objective, and the eco-development allows the adaptation of ideas to the local culture.

The international political conflicts on global sustainability stem mainly from competing approaches to development in both hemispheres. Chambers (1986) calls these: (1) environment oriented; (2) development oriented; (3) livelihood oriented, and he proposes merging all three as a basis for the national plans requested by the IUCN's Strategy. It is significant that eight years after the United Nations accepted the Strategy, only 40 nations of the 123 which support IUCN had prepared national conservation strategies.

If sustainability is to have a global future, many of the world's development and management institutions will need reforming. O'Riordan (1988) states:

> Sustainability is not regarded seriously by those who really count, namely those at the top of political structures and those who control the flows of national and international capital. The promotion of sustainability implies a reorganisation of agency alignments and priorities which cuts across the prejudices of those at the top. The full application of sustainability also demands new arrangements for budget-sharing and cross-organisational responsibilities that are deemed unacceptable or unworkable by those who benefit and operate through existing arrangements. By contrast, sustainable development, especially where it can be shown that thoughtless mismanagement of ecological systems can result in economic disaster and unprofitable capital investment, is now visualised as an appropriate way forward. But the pursuit of sustainable development is simply not enough.

If sustainability is not to be relegated to the category of 'seemed like a good idea at the time', it will need much greater commitment by North and South, each in their very different way. How aid and trade are handled will greatly affect the achievement of a stable globe. While it is heartening to have World Bank staff such as Robert Goodland (WRI, 1986) agree that the bank's major water schemes have caused 'massive damage', nationalistic governments have yet to be convinced on the finite nature and interconnectedness of our global resources. The depressingly feeble responses of many governments to IUCN and UNEP's calls to remodel development and aid programs generates little confidence that present goals in sustainability will be reached. Rather than allowing pessimism to smother the vision of a lasting world, what is required now is a redoubling of international efforts to convince governments of the original tenets of the global village. The education process must continue.

16.2 Government intervention — better than the marketplace?

Critics tend to generalise on the strengths and weaknesses of government intervention in agriculture, as if intervention per se is good or bad. There are valid reasons why many forms of intervention can encourage sustainable production. Such forms include the public funding of research, government extension services, special farm credit schemes, Rural Adjustment Schemes (RAS), marketing services and many other forms of government intervention. In some cases regulatory mechanisms control product prices, acreages under particular crops, export standards, storage and domestic sales.

The downside of such intervention may be stated as three shortcomings:

- The planning of many interventions is purely short-term and excludes environmental influences of the policies.
- Much of the intervention is parochial or regional and disregards regional differences in sustainability requirements such as slope, weather, erosion proneness and water use.
- Incentive structures too often mask ecological imbalances and over-use.

It is necessary to reduce, globally, incentives that cause the overproduction of subsidised food in developed economies, and to increase incentives for domestic food production in underdeveloped economies. Currently much of the Third World's productive effort goes into paying off international debt, while diet, health and family planning programs suffer. We have the technical knowledge to produce food by permanent systems and we understand the trade and aid framework required to maintain the resources. We lack the political will to agree on a global strategy for our common future. The agricultural systems developed over the past half century included the Green Revolution in which technology and high inputs contributed much to solving short-term hunger problems. These solutions were only fragments of what is now required in a shrinking world of burgeoning population. The greenhouse effect was not a factor in our earlier quest for growth, development and wealth. Today's world needs a *holistic* approach in which these three measures of progress are redefined. This is the new challenge.

Despite a long history of conflict on states' rights issues, in May 1992 the Commonwealth, the states and the Australian Local Government Association entered into a milestone agreement (the Intergovernmental Agreement on Environment 1992) which set out four important elements of sustainable land use in this country. For the first time in the nation's history a legally binding agreement was reached which:

- provided a cooperative nationwide approach to environmental policy and procedures;

- reduced the number and intensity of conflicts between the Commonwealth and the states on environment and conservation matters;
- gave greater confidence and security to government and private business decision-making on resource (mining, forestry etc.) and environment issues;
- produced the structures and procedures which, when applied in combination, would lead to better environmental protection.

The Agreement (IAE) clarifies the responsibilities and roles of each of the three tiers of government and spells out the principles on which all authorities should base their environmental policy. The IAE includes guidelines for integrating economic considerations, intergenerational equity, and the full costing of resources and environmental effects, and uses the precautionary principle. The nine schedules of the IAE cover:

- data collection and integration
- resource assessment, land use allocation and approval processes
- environmental impact assessment
- environmental protection measures
- climate change
- biological diversity
- national estate
- world heritage
- nature conservation.

Besides its national legal provisions, Australia is a signatory to the so-called 'Agenda 21' which is the 21st century plan agreed to by 176 nations at the United Nations Conference on Environment and Development in Rio de Janeiro in June 1992. Agenda 21 is a massive (700-page) document and considered, with some justification, to be the most important document in international history. Australia has prepared a report to the UN Commission on Sustainable Development showing everything it is doing to demonstrate its commitment to Agenda 21. As a basis for local action, Chapter 28 of the Agenda is crucially important. In Section 28.1 the Agenda (UNCED 1992) states: 'As the level of governance closest to the people, local authorities play a vital role in educating, mobilising and responding to the public to promote sustainable development'. This 'local agenda' has two objectives:

1 By 1996 most local authorities in each country should have undertaken a consultative process with their populations and achieved a consensus on a 'local Agenda 21' for the community.
2 All local authorities in each country should be encouraged to set up and monitor programs for ensuring that women and youth are represented in decision-making, planning and implementation.

The Commonwealth Government has funded a 'Local Agenda 21' campaign to help local authorities interpret and implement the Rio agreement. In view of Australia's reluctance to emphasise the role of local councils in environmental management, this is a welcome if belated change. Associated with the Local Agenda 21 is the so-called 'Local Approvals Review Program' (LARP), developed through the Federal Office of Local Government (OLG, 1992), to assist local councils with their approval procedures for subdivision and building. Sustainable land use should progress considerably more rapidly when the four principles of LARP are fully accepted and followed by all local authorities. The principles are that:

- approval processes should reflect efficient and effective decision-making for the *benefit of the wider community* (author's italics);
- state procedures should maximise the responsibility of local government in the approval process;
- local approvals should be accountable to the community;
- approvals should foster good developer/local government relationships.

Applying the last principle is probably easier said than done.

16.3 Market failure in environmental costing

One of the fundamental problems in attaining sustainable land use is the non-accounting for the so-called 'free goods' of the environment. Three main solutions are available but all require government intervention:

1. Legally enforceable regulations: These form official guidelines and standards which land managers are obliged to adhere to because non-compliance carries financial penalties.
2. Financial market incentives: Government uses these to add the full environmental cost to the price of land-based products. Examples outside agriculture include taxes on minerals and taxes on products that pollute — these may include all products requiring the consumption of fossil fuel.
3. Ownership rights to environmental goods: This gives property rights to individuals, organisations or governments over the amenity or service. Inputs are usually provided by the 'communal' environment. Giving private ownership rights on water, fish stocks or timber, for instance, usually discourages overuse because the resource now becomes an asset worth maintaining in good productive condition.

All three of these solutions have a place in Australia's ESD strategy and each can contribute toward accounting for the depreciation of our natural capital stocks, a cost which in Australia's national accounts must rightfully be subtracted from our gross national product (GNP). Nations such as Japan, France and Norway already use such accounting systems (Brundtland, 1988).

How do such fiscal policies relate to steady state economics? A steady state economy is one that 'maintains a constant stock of physical wealth and people' (Daly, 1980). Instead of attempting to maximise output (GNP), it includes a new concept of gross national cost

(GNC) which properly debits the country's economy with the full environmental cost of its production and consumption. The steady state of the national economy does not necessarily have to be kept at a constant level, for as technology changes so the sustainable level of the economy may change. The Worldwatch Institute (Brown, 1990) has even estimated the transition costs of converting the global economy to a 'sustainable earth economy' — $150B per annum, which was equivalent to 16% of the global military budget in the mid-1980s. It was estimated that the process would require this level of annual funding for 10 years to reach sustainability.

16.4 The Bargaining Model as a policy alternative

Environmental law is still in its infancy in Australia — the Environmental Law Association was not formed until 1980. Gardner (1989) has given a useful examination of the land conservation approach called the Bargaining Model, as an appropriate remediation technique for ecological problem-solving. In essence the law defines individual and group rights over natural resources which may be referred to as 'natural resource rights'. The Bargaining Model is defined as: 'the process whereby two or more parties attempt to settle what each shall give and take, or perform and receive, in a transaction between them' (Dorcey, 1986). In practice several important questions must be settled before the process can begin; for example, who should be included, how should negotiations be conducted, what information should be revealed, can a negotiated agreement be given legal effect? Negotiation and mutual agreement have long been part of contract law but multiple interests make environmental issues so complex that many established procedures are inadequate. The Bargaining Model, when functioning with the recognition of all affected interests, will tend to restrict ecological harm because the interests of those who gain from such harm are counteracted by the interests of those who suffer. This requires all affected parties to become directly involved in representative negotiation (Gardner, 1989).

A similar result may be obtained indirectly by some Environmental Impact Study procedures, but these give less opportunity for affected parties to be directly involved. The Bargaining Model may be useful in land use conflicts over rezoning, subdivision and environmental quality issues.

16.5 Land release schemes — WA case study

The question of why the rural industry is apparently immune from formal impact assessments is often posed in the Australian environmental debate. The soundness of decisions on the suitability of marginal land for permanent annual cropping has been repeatedly challenged, especially when salinity and erosion threaten regional stability. A case in point is the Western Australian government's release of 3 million ha (the size of two-thirds of Tasmania) in 1980, after such a scheme had been recommended by the state's Rural and Allied Industries Council (RAIC). A community study group (Pearce and Bradley, 1983) reported that: (a) the RAIC report was ecologically inadequate, (b) there was no public access to the Environmental Protection Authority (EPA) recommendations, (c) there was inadequate environmental data on the area, (d) there was no public involvement in the land alienation process, and (e) stability on adjacent farms was deteriorating. The sandy soils of the region were known to be particularly susceptible to both wind erosion and salinity and the chief of the Division of Resource Management had publicly declared 'It is quite clear that we have not developed an appropriate stable farming system for the south coast, nor indeed for many areas north of here' (Robertson, 1983). At the time 2 million ha of south coast soils were at risk from severe wind erosion and 250,000 ha of previous crop land had already been rendered useless by salinisation, which had risen 850% between 1962 and 1979 (Henschke, 1980). The study group called for a moratorium on land releases to allow for proper environmental assessment; full public participation in land allocation decisions; the redirection of resources to stabilise agricultural areas; and national deliberation on the environmental values (including threatened species) before large land release schemes are embarked on in future.

From a growth point of view, the Western Australian agricultural expansion must be regarded as remarkable by any standards, as shown by Table 16.1, which compares the situation after World War II with the recent past:

Table 16.1 Agricultural production: Western Australia

Statistic	*1948*	*1989*
Area cleared	6 0M ha	15.7M ha
Wheat production	0.9M t	5.2M t
Sheep numbers	10.0M	37 0M
Sown pasture	1.0M ha	7.1M ha

Source: Nelson, 1990

Today much of this land suffers not only severe salinity and wind erosion problems, but a range of other types of soil degradation such as structure breakdown, increasing acidity and water-repellence. While this prognosis appears pessimistic, research indicates that new technologies can solve many of the state's soil problems. Robertson (1990a) has analysed the ways in which Western Australian agriculture could benefit from better methods and lists the physical and biological factors that could improve it, as shown in Table 16.2.

Table 16.2 How Western Australia's agricultural problems can be overcome

• Soil structure improving.	• Less tillage. • No burning residues. • Better stock management.
• Soil acidity halted.	• Regular use of lime in fertilisers.
• Wind erosion reduced.	• More cover left on paddocks.
• Water erosion reduced.	• Contour working. • Less tillage. • Stubble seeding.
• Salinity rate of increase slowed.	• Better yielding crops. • Integrated farm management. • Revegetation of specific areas e.g. acid wodgils, deep grey sand areas, with trees and perennial forages.
• All saltland revegetated.	• Benefit/cost of halophyte recognised.
• Climate: warmer summer nights, more summer rain, drier winters, warmer winters.	• The greenhouse effect.
• Better crop yields.	• More timely planting. • Better varieties. • Better soil condition. • Better farmers.
• More varieties of economic plants.	• Availability of perennial plants. • Alternative crops. • Biotechnology. • New mindset.
• Local extinction negligible.	• Re-establishment and rehabilitation of remnant vegetation. • Catchment plans link up remnants. • Uneconomical farmland revegetated.

Source: Robertson, 1990

How does this case relate to land policy generally? It reflects the primacy of the old 'bigger is better' philosophy in state government planning circles, in which development criteria scarcely relate to stability and diversity. While state land policies have considerably improved in recent years, according to environmental criteria, making sustainability the main focus of policy is still to happen in many states. Recent land laws enacted in South Australia reflect the most enlightened policies to date.

Chapter 17

THE POLITICS OF LAND USE IN AUSTRALIAN AGRICULTURE

This final chapter deals with changes in the political powerplay in rural Australia and emphasises the decline in the national significance of the rural vote. The concept of 'country-mindedness'is examined in its modern context and the political value of conservation is examined in today's dynamic mix of conservation and employment in Australian politics. We examine the effect of political pressures on government support for the nationwide Landcare movement and how financial incentives can contribute to a more equitable cost-sharing in sustainable land use.

17.1 Country-mindedness — losing its clout

The trend toward a decreasing rural population has been continuing for several decades and by the year 2000 Australia will have less than 100,000 rural properties. Cockfield (1992) suggests that the fundamental difference between previous rural 'downturns' (recessions) and the present downturn is that in the past governments prevented any rationalisation of the rural sector; now the so-called market-liberalism is being allowed to take its course. Previous interventionist policies are being dismantled. Aitken (1988) has studied the concept of 'country-mindedness' as a major contributor to the agrarian myth which was instrumental in giving the rural lobby inordinate political power in the past. The idea that farmers form the backbone of the country has lost favour to the extent that the National Party (the Country Party of old) today commands no more than 9% of the primary vote. In 1951 the rural sector generated 26% of total gross product and 85% of exports (ABARE, 1992). Today it generates about 2.6% of total gross product and 20% of exports. Cockfield (1992) makes the important point that Australia cannot generate sufficient funds from the manufacturing sector to support (subsidise) the rural sector in the same way as the United

States and Europe do. Added to this uncompetitive position is the emerging trend toward more realistic pricing of natural resources such as irrigation water and timber from the state forests, with fees for previously free advisory services. The simultaneous tightening up of Drought Assistance policies, the scrapping of price support schemes (as in wool) and the Rural Adjustment Scheme which encourages further reduction in the number of producers, all combine to expose the rural sector to market forces for the first time in Australia. All these shifts in policy and intervention have a direct bearing on how rural land use, and agriculture in particular, will be practised and evaluated in the future.

Table 17.1 Farm sector economic and structural indicators

Gross value of farm production		(1990–91)*	$21,109M
Farm production as share of GDP		(1990–91)*	3%
Gross value of farm exports		(1990–91)*	$13,101M
Farm exports as share of total exports of goods and services		(1990–91)*	22%
Number of agricultural establishments		(1988–89)	126,551
Family farm ownership		(1985–86)	91%
Proportion of total population living in rural areas (including rural towns)		(1987)	30%
Farm employment		(1990–91)	407,000
Farm employment as a proportion of total employment		(1990–91)	5%
Land used for agriculture (hectares)		(1988–89)	467M
Proportion used for :	• cropping		4%
	• sown pastures		6%
	• extensive grazing		90%
Land ownership: (proportion of total land area)	• private	(1980)	13%
	• Crown lease or licence		54%
	• Crown other		33%
Livestock numbers:	• sheep	(1990–91)*	166M
	• cattle (beef)	(1990–91)*	22M
	• cattle (dairy)	(1990–91)*	2M
	• pigs	(1990–91)*	3M

* Preliminary
Source: Australian Bureau of Agricultural and Resource Economics, 1991

The socio-political position of the rural sector places landholders in a considerably weaker bargaining position than the one they previously enjoyed. The occupation, employment and production statistics for this sector have reflected a rapidly changing rural contribution to the national economy in the past three years, but Table 17.1 indicates the position up to 1991.

17.2 The development of rural conservation politics

The issue of land degradation has apparently been regarded, until recently, as both unimportant and unemotive by political parties, and in no way seems a vote-getter. This reflects a society which up to now has not placed a social stigma on rural land degradation. In fact, many rural community leaders participate in it. This situation is at odds with the politicisation of the more familiar environmental issues concerning rare animals, rainforests, coral reefs or wild rivers. Things beautiful, rare, green or furry have an obvious emotional appeal to most city voters and the current Labor government readily admits that its party's appeal to Australians' environmental awareness allowed Labor to win the previous election.

The fundamental problem of curbing degradation required impressing on the nation's leaders and decision-makers that this phenomenon, in all its varied forms, was jeopardising the agricultural future of the entire country. For decades state soil conservation authorities had vainly attempted to impress on their political masters the necessity of moving land protection higher up the political agenda. It had long been obvious that political parties were apparently quite prepared to accept that the agricultural industry was living off its land capital; that is, that present yields were produced at the expense of future soil productivity and landscape stability. Perhaps even this contention credits politicians with a greater insight into the issue than is justified, but the outcome remains the same — erosion and salinity simply did not become political issues until a few decades ago. Starting in the 1940s, political interest in erosion and soil losses gained momentum until the 1950s. This represents the first wave of concern about land degradation, analysed by Stadler (1989), and was followed by a period of disinterest by political parties which lasted until 1983 when the Federal Government first entered the land degradation arena seriously. Since that time, there have been substantial changes in political interest in the problem, such as the Prime Minister's personal conversion to identifying land degradation as 'our nation's major environmental problem'.

In the face of well-publicised irrefutable evidence of the economic implications of soil erosion and salinity, the Federal Government in the early 1980s could no longer ignore the problem. Minister Kerin sought to initiate Commonwealth involvement in national land improvement with his Soil Conservation (Financial Assistance) Act of 1985, and $1 million was allocated to a problem which was costing an estimated $600M per annum. This level of funding reflected Treasury's well-known reluctance to recommend appropriate funds to what had long been regarded by the 'fiscal fiends' (Kerin's terminology) as a bad investment of public funds. Although a 1978 national study had shown that 51% of Australia's farm and pastoral lands had suffered degradation, the problem had been either ignored or relegated to the 'too hard' basket of election issues by successive governments. Politicians apparently regarded the problem as very expensive and possibly non-reversible, with very little political mileage in identifying with this rather boring and unemotive cause. However, in a geographically patchy way, from the early 1980s researchers and leading landholders began to pay heed to pressure from the conservation lobby to act against rural environmental decline. Politically speaking, the 'big green finger' pointing at the farmers slowly but surely became a new issue for the Australian electorate and its representatives.

An important element of Australian rural politics is the assumption made by various user groups about the primacy of their particular type of land use (e.g. grazing, cropping, timber). As the population increases (presently at 1.4%, compared to the world average of 1.4% in 1993) the competition for land becomes more intense. Australia is in the favoured position of still having vast undeveloped areas in a crowded world. One such 'open' area is the top end of Queensland where the Cape York Peninsula Land Use Strategy (CYPLUS) presents a case study (Holden, 1992) in how attitudes to land use alternatives have changed — at least among the political power-players.

The CYPLUS program was set up by the Commonwealth and Queensland governments in 1991. In 1992 the Natural Resources Analysis for the Peninsula was initiated, aiming at sustainable land use and involving the community in the planning process. The community is represented by the Cape York Land Council, the Torres Strait Island Coordinating Council, the Aboriginal Development Association, the Cape York Peninsula Pastoral Advisory Group and the Cairns and Far North Environment Centre. The first community workshop made a list of information needs, then considered the best ways of collecting the resource data. It was decided to form community working groups. The program was coordinated by a quality control group to monitor how the data collection met the requirements for sustainable development. Another committee was established to deal specifically with indigenous interests. Data was categorised into general topics: nature, people and land (environment, social issues and land matters). The database includes present land use, tenure, coastal access, current industries, heritage sites, population distribution, parks and their management, and the underpinning values in each land zone.

Cape York represents an area that can be planned almost *de novo* because of the current low level of development. In this case the interests of Aboriginal people and of tourism are likely to be given preference over the traditional interests of grazing. The rural politics of the Peninsula are now dominated by latterday interests such as coastal tourist condominiums, a spaceport and the exploitation of silica sand. A study of land tenure in this region shows that genuine cattle producers as well as bogus 'pastoral companies' hold vast tracts of land, including the Starkey lease recently resumed by the Queensland government.

17.3 Party political support for conservative land management

Traditionally Australian political parties have represented workers (Labor), business (Liberal) and landholders (Country/National). While such divisions are simplistic, they help to explain why Labor and Liberal never acknowledged land degradation as being even remotely connected to their 'true believers'. The Country Party knew about erosion and salinity but responded to all land crises in terms of support for the rural people, rather than by policy change to encourage sound land management. In such a political framework it is to be expected that 'looking after the voters' far outweighed the need to look after the land. In the absence of a national education program to gain acceptance of stewardship, 'band-aid' policies would remain.

At state level there had been ad hoc legislation to cope with wind erosion since 1923 (South Australia) and with water erosion in Western Australia (1945), Queensland (1951), Victoria (1958), Australian Capital Territory (1960) and Northern Territory (1969). Soil conservation authorities were set up in most states and government extension officers were available to advise landholders on degradation problems. Despite this belated recognition of the problems, few state politicians ever used soil conservation as a major plank in their election campaigns. A notable exception was Bill Scully in New South Wales in the mid 1940s. In more recent years parliamentarians from all states and in both Houses of Parliament spoke with increasing fervour on the now accepted issue of land degradation.

Labor's political base had long been in the cities while the Country Party continued to convey an image of the fading squattocracy. Despite these differences in political background, both political sides claimed to represent the ill-defined 'little battler'. In the case of the little rural battler, serious problems of small farms on the old soldier settler schemes, saline irrigation developments and drought-prone arid grazing regions all led to policies which avoided land capability problems, rather than facing and solving them. Bradsen (1988) has extensively reviewed past and present soil conservation legislation and has found it inadequate.

Neither the socialistic ideology of Labor nor the free enterprise emphasis of the Country/National party did anything to stop unsustainable land management practices. This is not unexpected in a society that expects government to look after it in the short term, even if social vision invariably gives way to political convenience. In this situation it is not surprising that statesmanship gave way to pork-barrelling of the most expedient type. Once again the three year term-of-office was given priority over the long-term sustainability of the land.

17.4 Political ideology and Landcare's future

The Landcare movement was born out of a belated political recognition of the financial implications of not acting against loss of production potential. It is probably true, however, that the government would not have backed this community phenomenon if the Labor Party had not been confident of scoring votes by appealing to the environmental sympathies of the urban (87%) majority. In other words, electioneering in the late 1980s could safely emphasise Landcare policies in a political climate that almost guaranteed that such policies would capture the swinging voters. Landcare, as an environmental issue, is very different from causes such as the 'green bans' which Jack Mundy and the urban Labor followers of the early 1970s had supported. Landcare is primarily about farmland, while virtually all of Labor's previous environmental initiatives had been about city-based 'greenie' concerns.

The fundamental political question on Landcare is 'Who should pay for land rehabilitation?'. When the problem is generally attributed to private landholders in their quest for personal benefit, financing remedial action becomes politically sensitive. Answering this question requires government to clarify its stance on land stewardship as a serious goal of rural policy. Putting it another way, the national success of Landcare depends on landholders adopting a land ethic according to which they manage the land in the fashion of stewards of the community's land resources. Politically, it follows that if individuals are expected to become 'guardians of the common resource', they should be supported by the beneficiaries of this noble task. This requires a fiscal policy that gives sufficient incentives to take on a custodian role without losing personal benefits. Government policy which rewards the custodians of the nation's wealth-generating resources could be seen as 'politically correct' in such circumstances.

Positive rural conservation incentives are a far cry from the negative policies of earlier years. Right up to 1983 practices such as tree clearing had been fully deductible expenses for income tax purposes. Similarly, the drought assistance rorts encouraged overstocking in much of Australia's most fragile pastoral country until the mid 1980s. Drought policy is perhaps the most striking example of inappropriate policy — it seemed to reward bad management and encourage environmentally degrading behaviour at the expense of the taxpayer. Blyth and Kirby (1985) have explained the effects of such political shortsightedness on Australian landscapes. More recently Boer (1990) has reviewed the limits to individuals' property rights.

Landcare is a landholder-driven movement financed by the federal government and carried out by state advisory staff. It is coordinated by federally funded officers who help to keep community trials and demonstrations accountable. Some 2,200 Landcare groups, backed by a federal allocation of $340M for the Decade of Landcare (ending in the year 2000), form the organisational structure of

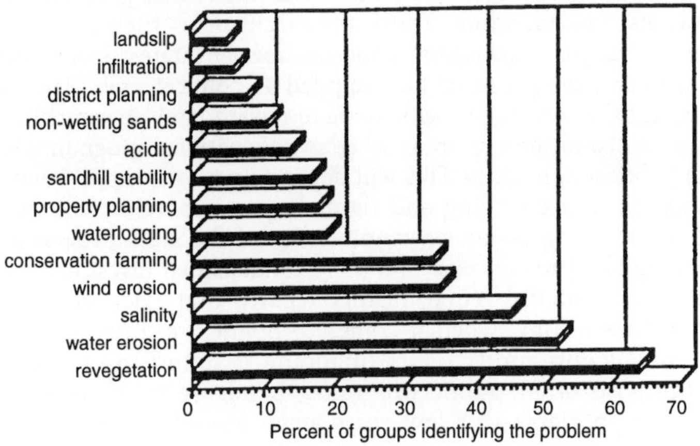

Figure 17.1 Issues identified by Landcare groups in South Australia
Campbell, 1992

Landcare. New tax reforms (Section 75D) provide for a wider range of deductions for land rehabilitation activities, and new drought policies tend to benefit pastoralists who stock their land more conservatively. However, despite the progress made with more enlightened political and fiscal policy, government continues to undervalue Landcare as a national necessity. The percentage of GDP spent on securing the nation's export-generating resources is too low by at least half. The delays and superficiality of political attempts to make real progress with ecological sustainable development reflect a persistent lack of political will to give land degradation the attention and resources it warrants as a national priority.

Although Landcare has been hailed as an important and unique home-grown model for national action at the Rio '92 Global Forum and elsewhere, the real meaning of land stewardship has yet to penetrate the political engine driving this most urbanised young nation. Converting noble sentiments and catchy slogans into behaviour of the lasting kind will require a considerably greater shift in ecopolitical thinking than has occurred to date. Without a more appropriate policy framework and reward system, noble causes and great works by individuals cannot be expected to overcome two centuries of neglect. As an example of the work required, Figure 17.1 shows the issues identified by Landcare in South Australia.

Many uncertainties lie between now and the end of the Decade of Landcare. First, it should be recorded that it took only one term of office for Labor to drop 'the environment' as an election issue — it didn't even rate a mention in the 1993 campaign. Second, there is an increasing conflict between federal and state governments over the financing of Landcare support staff. While states are responsible for land use, some are reluctant to fund the coordinators and facilitators of the multi-faceted Landcare program in case the federal government withdraws its direct funding. Third, the mechanisms such as cross-compliance, income equalisation deposits, special interest loans and rent reductions have yet to be fully applied as political tools. This is not to suggest that nothing is being done to reward land stewardship.

Crucial to Landcare's future is whether this national program will retain its political popularity. This will depend on what priority voters give Landcare in a changing and tightening economy. It will also depend on the progress and accountability of the state programs. While it is difficult to conceive of a more fundamental investment of national consolidated revenue, history records that election expediency wins every time when it comes to political priority-setting. Even if the accountability of its financing is found to compare favourably with other public programs, Landcare may easily be deemed unsupportable in a few years when its political gloss has faded.

Landcare must be accepted as a permanent national program, just as education, health and social services are. Commonwealth authorities have repeatedly emphasised that federal funds are catalytic only — that is, they should be regarded as 'seed funding' only, for initiating but not maintaining or expanding trials and demonstrations as a basis for developing a nationally accepted land ethic. A combined educational program for voters and politicians is the only way of securing a future for land stewardship in Australia. Policies that balance voluntary and regulatory incentives in a 'soft sticks and hard carrots' approach can make Landcare work even in our kind of democracy. The four pillars of success identified by Campbell as the foundation for Landcare to achieve its goals are shown in Figure 17.2.

In the final analysis of Landcare's future as a national movement,

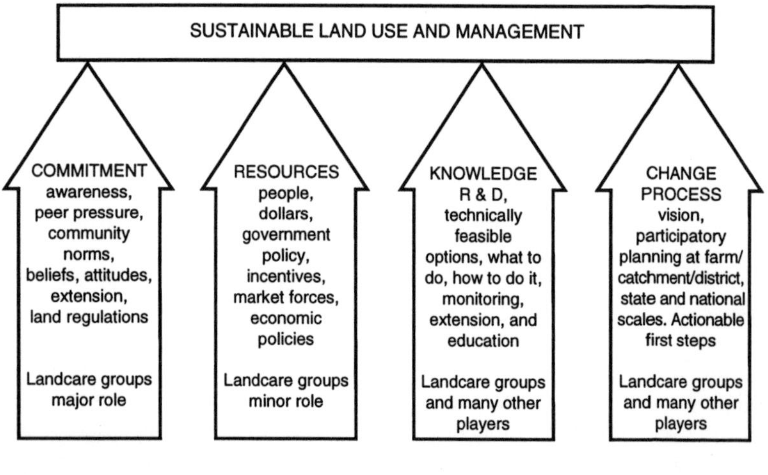

Figure 17.2 Four requirements for progress in the implementation of Landcare
Campbell, 1992

the political climate in years to come will directly influence funding and other support. Australian political thought will be subject to the same optimism/pessimism dichotomy of opinion as that which influences government decisions worldwide. The extent and dynamics of various global resources and their interrelations are such that even analysts with similar information take different views on the future of resource sufficiency. Kahn, Brown and Martel (1976) encapsulate four broad views that form the basis of analysts' optimism/pessimism, using the pie analogy:

A *Convinced neo-Malthusians* assume that renewable resources can be estimated accurately enough to show that resources will run out, and that many critical resources will run out in the next 50 years; the resource pie shrinks and economic growth, which makes the rich richer, can only make the poor poorer.

B *Guarded pessimists* recognise resources as an 'uncertain' pie in which traditional and replacement resources are fraught with insecurity. Over-conservation and excessive consumption must both be guarded against.

C *Guarded optimists* have faith that past economic and technological growth can increase production, and that higher consumption by the developed nations can benefit all people. The 'best bet' and 'least risk' solutions are assumed to include rapid technical and economic development.

D *Technology and growth enthusiasts* consider the pie unlimited which means that capital, technology and educated people are the most important resources for an optimistic future. No real limits to growth are recognised, except for population in some developing nations. The constraints of a finite earth are offset by further exploration, expansion and conversion. Humankind is seen as rising above the dire predictions of the doomsayers.

These philosophical options are no less real for Australians, whose vast space, high material standards of living and relatively small population favour the optimism of the Technology and Growth Enthusiasts. Such optimism may be justified by a population growth rate of 1.5% in 1991 and 1.4% in 1992, and immigration down from 140,000 in the 1970s to 102,000 in 1992. Landcare will need population goals if our planning of resource use is to be realistic and not in conflict with other social goals.

17.5 Costing and pricing land degradation

It has become common practice to divide the damage caused by land degradation, notably water erosion, into onsite and offsite costs, which can be subdivided into private costs (borne by the landholder) and social or public costs (borne by society). Economists may also distinguish priced and unpriced degradation effects where costs can be allocated only to certain tangible damage. Yapp (1989) suggests that in general terms the value of natural resources is determined by what people are willing to pay (WTP). Where direct pricing is problematic, alternative pricing mechanisms can be applied. One such alternative is the so-called 'contingent valuation' based on a hypothetical market;for example, willingness to accept variable compensation. Another pricing alternative, termed the 'surrogate market value', relates the resource or its services to another associated market situation. Finally there is 'hedonic' pricing which estimates the cost of market value lost by land degradation. Whichever method is used to price the effects of degradation, it is still hard to show that money spent on degradation control (Landcare) is a good investment in terms of outcomes. For instance, Yapp (1989) points out that in New South Wales soil problems such as structural decline, acidification and salinisation are growing in terms of conditions and in economic terms, despite the expenditure of $32M annually for the decade ending 1987. Similarly, in the United States McConnell (1983) states that despite annual expenditure of about $300M per annum for nearly five decades, 'there is little evidence that erosion has been curtailed'.

The full cost (including long-term social cost) of resources is often referred to as the single most important change required in Australian environmental policy. In combination with education and regulation, financial incentive holds the greatest potential for advancing environmental awareness to sustainable resource management. The Canadian Government (1992), which has evaluated the benefits of applying 'economic instruments' (financial incentives) to environmental programs, suggests at least five ways in which environmentally appropriate land use programs can be encouraged:

- Resource policy can offer cost savings in return for compliance with set standards of environmental management. In other cases cross-compliance can offer financial assistance on condition that certain management criteria are met.
- Full resource pricing based on the total socio-economic value of the resources used offers real incentives to use resources without waste (e.g. water) or exploitation (e.g. forests).
- If financial incentives make it cheaper for resource users to change their methods and technology than to carry out imposed resource changes, businesses will improve their methods more rapidly.
- Policies that raise revenue from disincentives with reduced damaging practices can provide for the revenue-generating instruments to fund further awareness programs, expanded enforcement, and better management training in resource use.
- Well-designed economic incentives can greatly increase the rate of congruence of private companies' environmental and economic goals. Specific rewards for stewardship activity lead to a recognition that environmental 'best practice' is integral to, rather than peripheral to, long-term family and company business ventures.

Although space does not allow for a detailed analysis of what has become known as 'environmental economics', a brief outline of the criteria for selecting economic instruments (to encourage sustainable development) is warranted here. The Organisation for Economic Cooperation and Development (OECD, 1991) has identified five ways to measure how well such instruments perform: environmental effectiveness (does it produce planned ecological outcomes?); economic efficiency (does it give value for money?); equity (does it balance private benefit with public good?); administrative feasibility (can it be applied at reasonable cost?); and acceptability (do stakeholders consider it appropriate?).

17.6 Tax incentives for sustainable land use

Over the years each state and the federal government have developed a range of incentives to encourage better land management. Some of Australia's past land policies have actually rewarded practices such as overstocking and injudicious clearing by offering ill-conceived tax deductions and financial assistance. Most of these disincentives to good land management have been removed in recent years and replaced with more effective incentive policies. The details of financial incentives vary widely among the states, but the federal government modified the Income Tax Assessment Act 1936 (Section 75D) in 1990 to expand considerably the scope of Landcare taxation arrangements. The current wording of the guide to the Act is as follows:

> Primary producers and other businesses earning income from the use of rural land are entitled to claim an outright deduction (100% write-off in the year of expenditure) for certain capital expenditure, if the principal purpose of the expenditure is for one of the landcare operations listed in the section. Expenditure on the following operations qualifies for outright deduction, as specified under the following paragraphs of subsection 75D(1B):
>
> (a) the eradication or extermination of animal or vegetable pests from the land
> (b) the destruction of weed or plant growth detrimental to the land
> (c) measures for preventing or combating land degradation, other than by the erection of fences on the land (because erection of fences is covered by paragraphs (d) and (e) of the Act)
>
> ▲ this could include, for example, soil conservation earthworks such as drop structures and contour banks, planting trees and shrubs to control salinity or stabilise erosion gullies, establishing windbreaks to prevent soil erosion, and regenerating native vegetation
>
> (d) the erection of fences (including any alteration, extension or addition to fences) on the land to exclude livestock or vermin from areas affected by degradation in order to prevent degradation getting worse in those areas and to assist in reclamation of those areas
>
> ▲ this could include, for example, the cost of fences erected to keep stock out of erosion gullies or salt discharge areas; or the cost of fencing off trees and shrubs as part of an operation to prevent or combat land degradation. Note: Land management plans are not required for this deduction.
>
> (e) the erection of fences (including any alteration, extension or addition to

fences), which separate different land classes in order to prevent land degradation in accordance with an approved land management plan

(f) the construction on the land of levee banks or similar improvements having like uses

(g) the construction on the land of surface drainage works or subsurface drainage works (excluding the drainage of swamps or low-lying land) for the purpose of controlling salinity or assisting in drainage control

▲ this could include, for example: land levelling and grading and contour banking as an integral part of the process of combating or preventing the problems of soil erosion, land salinity and drainage; drainage channels and associated earthworks and culverts; and subsurface drainage such as installing drainage bores and laying surface and subsurface pipelines, constructed to combat land degradation.

Any person, including primary producers and rural business people themselves, can prepare the land management plan required for fencing to prevent land degradation under section 75D(1B)(e). But, because the plan must be approved by either an authorised officer of a State or Territory land conservation agency or an authorised farm consultant, landholders may wish to obtain advice in identifying or confirming the land classes on their properties.

Another aspect of taxation is the introduction of Income Tax Equalisation Deposits (IEDs) which assist primary producers to pay less tax in years of low production. The current income averaging system has the disadvantage of confronting producers with a sizeable tax bill when they can least afford it, except when they actually make a loss. Although income averaging is much fairer for people with fluctuating incomes than a straight tax on each year's income, it has a drawback — it tends to put financial pressure on producers when they're in hard times. To alleviate this pressure, the National Farmers' Federation has proposed a plan which allows producers to carry forward tax credits for unused parts of marginal tax rate brackets in the same way as they can currently carry forward losses to be offset against the next year's income. The scheme, Carry Forward of Tax Credits (CTFC), would allow producers to pay the same total tax over a period of years as someone using income averaging or on a regular income. The difference is that the timing of tax payments under CTFC is more appropriate for producers working under variable weather conditions (Goucher, 1994). The proposed system would require credits to be used as early as possible and would put a limit of 10 times the size of an income bracket (e.g. $153,000), on the total credits that could be accumulated in each bracket.

In attempting to put taxation in a planning framework for sustainable development, Young (1993) has proposed a set of rules whereby the complex concepts of inter-generational equity and the precautionary principle can be accommodated. Although these general rules would have to be refined for local situations, they offer regional planners a better set of guidelines than others used previously. The equity and caution objectives are achieved with four rules:

Rule 1. Maintain an efficient, diversified and ecologically sustainable economy. This rule is based on the assumption that the richer the current generation, the greater the possibility that this generation will be generous to its successor.

Rule 2. Maintain the natural capital (ecosystems and resources) that has no substitutes, and provides valuable services to the community. The assumption is that any depletion of this 'capital' could have irreversible and/or adverse consequences for coming generations. Natural capital is necessary to maintain ecological functions, to assimilate wastes and to supply food.

Rule 3. Adopt the precautionary principle as a basis for responsible action in uncertain conditions. This principle ensures that policy-makers and decision-takers take special care with risky alternatives when they have insufficient scientific knowledge to be confident of the outcomes. The principle includes an onus on developers to prove that no adverse consequences will threaten this generation nor its successors.

Rule 4. Continue to take decisions after the best evidence on irreversible damage from alternative actions has been evaluated. This rule aims to avoid indecision and delays while eliminating long-term environmental costs.

Rule 5. Increase the diversity of social, economic and environmental elements of national wealth so that the next generation (at least) inherits an equivalent endowment of resources. This rule constrains economic development to be consistent with the enjoyment and consumption of natural capital by later generations.

To apply these rules to specific policy for sustainable agriculture, we can return first to Solow's (1974) near-classic exposition, *Intergenerational Equity and Exhaustible Resources* which initiated the current debate on sustainability from an economic perspective, and led to the identification of the four basic properties of efficient agro-ecosystems: productivity, stability, sustainability (akin to resilience) and equitability (Conway, 1987).

In an attempt to evaluate alternative policies designed to encourage Landcare and sustainable farming, Black (1994) has distilled the work of several policy analysts into a set of 10 criteria for appraising land use policy instruments. These instruments are grouped into educational, regulatory, fiscal and institutional (redefining resource rights). Consider first the regulatory approach. Regulations backed up with financial penalties are akin to taxes that are paid beyond a set limit of resource use; such regulations require monitoring on a vast scale. This can be done by satellite (Epps and Crittenden, 1992) and remote sensing. A 1993 survey indicated that 57% of Australian landholders favoured greater use of such technology to identify degradation on private properties. The survey also showed that 55% of producers supported the notion of zoning marginal land into zones of permitted agricultural intensity (Reeve and Black, 1993).

Fiscal measures usually consist of taxes and charges to discourage certain practices, or consist of subsidies to encourage other practices. Subsidies may be linked to some cross-compliance arrangement, as referred to elsewhere. The complexity of administering cross-compliance has made it less popular than other approaches. The effects of the concessions and subsidies on desirable practices can also be gained by charging levies on degrading practices. As explained by Black (1994), such levies should be set at a level equal to the cost imposed on society by undesirable practices. This 'polluter pays' principle recuperates the total social cost of pollution from those that cause it. Since identifying the polluter is often very difficult in agriculture, several Scandinavian countries charge levies on all polluting substances such as pesticides and fertilisers, to provide funds for water quality control programs.

One aspect of policy not widely used in Australia to encourage sustainable farming, is what Young (1992b) has termed 'sustainable investment' in well-designed 'resource-right' systems. Such systems are economically efficient because they use renewable resources that are resilient when exploited, and are allocated their most highly valued form of use. Such systems do not require changes in government policy to seek new markets, apply new techniques or incorporate the latest information. Young maintains that such resource-right systems can compensate those who give up certain resource rights and reward those whose management benefits the environment.

Australia's choice of policy instruments to encourage sustainable rural production is complicated by the fact that most agricultural products are sold on world markets which are also supplied by countries less concerned with sustainability and more concerned with competitive advantage in the short term. Only an international convention can overcome ecological inequality caused by such pressures.

EPILOGUE: THE ROAD AHEAD

New Values

For our perspective ahead, we must first look back — to when modern attitudes to the environment began to change. The days of unquestioning acceptance of growth and development as inherently good attributes of a 'progressive' society came to an end in the 1970s when the focus on uranium mining, Aboriginal land rights and the preservation of wilderness challenged the values and the conscience of this nation for the first time in its history. The idea that development and the science and technology that support it have only positive outcomes changed at that point in our history. Dominating nature and regarding it only as a resource for generating wealth have given way to alternative but equally valid world views and values, each contributing to the conservation ideal as shown in the figure below.

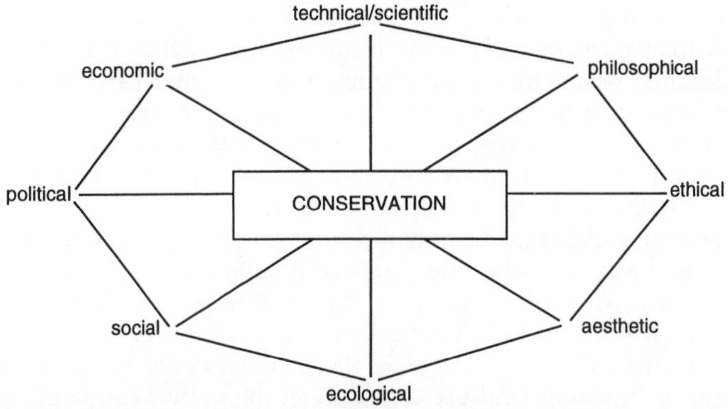

Complementary approaches to conservation
O'Riordan, 1971

Lines (1991) suggests that Australia's real choice does not lie between different market mechanisms or alternative fiscal policies, but between industrial capitalism and collective social action. This means a choice between destruction and a more sensitively and democratically planned future that does not exploit the earth. This road ahead is also the road to a real freedom and a lasting social justice. Why? Because it is based on a caring and sharing society in which quality of life comes not from material acquisition but from the sensitive relations between humans themselves, and with nature; a sustainable society in which resources, population and pollution are kept in a lasting equilibrium. The selective-protectionist approach of well-endowed 'green' organisations must give way to a genuine change of heart over the concept of success in Australian society. New success, new heroes and role models must form the foundation for the New Australia. An acceptance, indeed a reverence, for steady-state economics is the catalyst for this life-giving change in our national character. This attitudinal change can only be achieved nationwide by taking active steps toward designing, establishing and persevering with a diverse but comprehensive education program that combines ecological understanding with the values of a sustainable society. Because the western tradition in the past showed little understanding of these new values (indeed, conflicted with them), this change may take at least a generation to achieve. If new awareness leads to new knowledge, and if new knowledge leads to new attitudes, then new values can develop from this sequence of social changes.

Whatever the historians would have us believe about how the landscape moulded our forebears' actions and behaviour, it was the Christian-Judeo values of right and wrong, reinforced by the Anglo-Saxon concepts of progress from the Industrial Revolution, which engendered the unfeeling, uncaring, unsympathetic attitude to nature — one that damaged more land per capita in two centuries than anything else in this planet's history. Australia's record is one of robbing nature's treasure store, and if the future is to be different it will be either because we are intelligent enough to read nature's early warning signals (and some not so early) or because nature throws us off. The threat of ecosystem collapse has yet to penetrate the political and economic mind of this nation. We are slow learners, despite the historic lessons of failed civilisations, so plain for all to see.

The road ahead must be one of reconciliation, a process in which racial harmony is only one part — perhaps the minor part. Reconciliation with nature in a pact of mutual respect, of 'live and let live', is the key to survival. Australia can survive without many of the programs currently promulgated in the national interest, but unless we grasp the inherent land use capability of the Great Southland, our national house is 'built upon the sand'.

The road ahead must also be charted by individuals and organisations who can put aside the social and political norms of the day and reject the superficial values of a short-term parliamentary system that does nothing to encourage vision and statesmanship beyond expediency and the appeasement of voters. The orthodox society must give way to the sustainable society inspired by the spirit of 'doing unto others ...' — which will in future include non-human 'others'. In such a society, agriculture and land use will be much more diverse, more organic, more energy-efficient and more resilient because of its broad base, its flexibility, its options and its natural equilibrium.

The future Australian society may have a check-list of basic human needs, and if these can be met by permanent systems, the road ahead looks bright indeed. These basic needs may be grouped first into the fundamental physical needs of clean air, clean water, nutritious and varied foods, adequate housing and energy supply. Arising from the physical needs is the need for adequate waste disposal systems. The second group of needs encompasses social conditions such as varied employment and the intellectual nourishment of education. Associated with this group are needs that are still luxuries in much of the Third World: the physical needs for recreation, mobility, a satisfying environment, health, welfare and safety. The final group of needs for a well-rounded sustainable society includes the fundamental values: freedom of body, mind and spirit, privacy, personal communication and satisfying human relationships in a balanced, heterogeneous community.

Can Australia attain this Utopian goal for its citizens of the twenty-first century? The answer can only be given in terms that relate to other nations. All things considered — physical, economic, social and environmental — Australia seems to have the best chance of developing a global case study in sustainable societies.

New Education

Teaching the merits of no-growth economics in the early 1970s put innovators like Daly and Boulding in the firing line of all those who worshipped at the Keyensian altar of supply and demand where the marketplace drives the system. The centrality of population control has gained international attention since the global family reached 5 billion in 1990, and the conditions of the 'haves' and 'have nots' are now central to the steady-state debate where Third Worlders demand the attainment of a higher state before it is declared steady.

From an educational point of view, aesthetic appreciation of the environment requires both knowledge and feeling — *intellect* and *emotion*. When our value-free curricula became a market of ideas for students to choose from, the 'should/ought' element of ethical behaviour lost out to the free thinkers. Free thought has much to recommend it, but in this era of greedy consumerism we shall require substantially new thinking if we're to survive as a planet-bound species, and the religion of economics will have to be replaced with something like Tyler Miller's 'earthmanship' if our fragile ecological systems are to continue sustaining us.

The recent national report on Australian agricultural education is disappointingly narrow in its scope, and shallow in its consideration of the links between technology and the formation of fundamental social values. Can such values be nurtured by university programs like Harvard's 'Moral Reasoning and Social Analysis'? If so, is it the university's role to help students understand 'where values come from and how they are shaped and kept' as suggested by Borsh of Princeton? In an inaugural address of 1971 the writer's appeal included three challenges to the university in South Africa:

1 To develop an ecological conscience by rethinking man's place in nature.
2 To extend ethics to include responsibility for the environment.
3 To expand the idea of human rights to include rights to life of non-humans.

Twenty years later, these challenges remain valid, but are more urgent, in Australia.

Our education system has had little effect in reining in the technological imperative which says '*because we can, we do*'. Our ability to intervene in natural processes and our newfound desire to improve on nature have given technologists free rein, as if '*can = must*' and '*able = ought*'. Teaching ethical values requires the wisdom to distinguish which of our new abilities will contribute to a sustainable future — in essence, to replace cleverness with wisdom.

As part of our educational perspective we must ask ourselves whether Australia is ready to include a land ethic in our community values. For three decades after Leopold (father of the white man's land ethic) put forward his ideas, isolated Australians called for the application of his philosophy in Australia. In later years leaders in Australian thinking such as Passmore, Routley and Birch urged Australians to not only become more sensitive to their environment but to develop a respectful partnership with their unique fauna and flora. Even more recently the emergence of the Landcare Movement in 1988 focused attention on the need for a new belief-system as it applied to human/land relations. Australia's first paper on this subject — 'Land Ethics: A Necessary Addition to Australian Values' (Roberts, 1984) — received a cool reception at a time when the population was polarised between producers and conservationists.

The moralists among us question whether we and our national leaders know what values we stand for any more; with all our talk of 'values clarification' in education, have our institutions forgotten to teach the specifics of what is right for our future society? Has our emphasis on the importance of having choices led us to maintain such an open mind that we don't close it on anything? Has this become our only basis for personal values and action?

Earlier thinkers such as Plato, Locke and Rousseau remain important not because they uttered eternal truths, but because they showed us how to seek such truths from our own experience. The modern issues of population, resources and pollution combine to force us to heed a new group of truth-seekers — Ehrlich, Commoner, Hardin, Daly, Capra, Leopold and Suzuki — and closer to home, Birch, Passmore, Dawe and Oodgeroo Noonuccal, all of whom urge us to understand our place in nature.

In reminding us that the greatness of civilisations is not usually measured in industrial productivity or balance of payments, David Myers recently revolted against the current technological fix with an unexpected outburst: 'To hell with producing a Clever Australia! What about producing an Artistic, Literate, Spiritual and Ecological Australia?' Perhaps Myers is intimating that our unfettered cleverness has progressed without concomitant wisdom, has damaged our global habitat, and has put sustainability high on today's agenda.

Much of our problem in science has been our preoccupation with analysis at the expense of synthesis. This has led us to study sterile fragments of the real problem which doesn't break down easily into the bite-sized chunks required for a neatly packaged thesis. The move to integrated holistic thinking has come none too early to some of our learned faculties. Other important goals in education include developing not only knowledge but interest, abilities, understandings, attitudes and ideals. Knowledge is a means to these ends. The intellect and the emotions must both converge on new knowledge to develop our commitment to worthy causes. Intellect and knowledge are insufficient on their own.

Few of us could have missed the current drive for the Super Highway which will link all the electronic media channels to give us more and faster information than the human brain has ever received. Perhaps it has occurred to you, as it did to me, that the major problems of humankind are not caused by lack of information, but rather by the lack of meaning and values underpinning our present information overload. One of the outcomes of excess information has been termed 'Edutainment', a value-free albeit interesting conglomeration of information that entertains and occupies people without actually educating them. As a former agricultural colleague of mine observed, 'Students should not spend their time collecting spades, rather they should be taught how to dig' — for relevance, meaning, truth and underlying values. There is probably wide agreement that the university should teach good thinking — critical, original and independent — not what to think, but how to think, not in a value-free system but recognising the harsh realities of long-term survival.

It was probably the unusual prosperity of the 1960s which, through increased income, led to virtually all (white) Australians becoming middle class. With this 'middle classness' came a respectability which avoided participation in public conflict, notably environmental conflict. Now this nation has been plunged into an era of unprecedented change, not just cultural, economic and technical, but also in fundamental values — marriage, employment, gender roles, Asian relations and even the constitution. Will Australia change from a collection of inherited cultures to a new, more mature nation, with a diverse but cohesive identity? Will the so-called Anxious Eighties lead to clarity, certainty and stability, or will the increasingly obvious Last Straw Syndrome continue, in which small things cause us to snap, as excessive change overwhelms us? Much depends on how our young people define success and whether we can accept that our economic imperative must change forever, as John Naisbitt predicted in his best-selling *Megatrends* some years ago.

New Economics

The road ahead for Australia will lead nowhere until values, attitudes and education are supported by economic policies and regulations which include environmental audits and ecological costing. This means comprehensive accounting of what Barbier (1991) terms '*total economic value*', which includes 'use values' as well as 'non-use values'. Young (1992a) lists examples of the Use Values as follows:

1 Direct Use Values — timber, crops, recreation.
2 Ecological Function Value — flood control, nutrient cycling, climatic effects, photosynthesis.
3 Option Values — future genes, drugs, substitutes for depleted resources.

The Non-Use Values are:

1 Existence Value — the knowledge that the resource exists.
2 Bequest Value — the benefits to future generations.

So Total Economic Value is the sum of the above five values. If economic policy is to value resources properly then each of the above five values must be computed as subsets of the total value and then factored into the national balance sheet. Young (1992a) contends that in any environmentally costed economy the following pricing formula should operate:

Efficient price for a resource=
marginal cost of supplying the resource to the user
+ marginal cost of any lost ecological functions
+ marginal cost of any pollution that the resource use imposes on other people
+ marginal cost of lost future options
+ marginal cost of lost existence and bequest values

The mechanisms by which dollar values can be assigned to each of these cost functions have not yet been fully developed. However, the complexities of the calculations and the possible lack of precision of the estimates do not justify abandoning the procedure in favour of inappropriate traditional economic costings.

In allocating costs to particular operatives, the much-quoted 'Polluter Pays' principle is likely to play a major role in future Australian economic policy. This principle, developed twenty years ago, is explained by its originators as follows:

> The Principle means that the polluter bears the expense (of pollution prevention and control) decided by public authorities to ensure that the environment is in an acceptable state ... Such measures should not be accompanied by subsidies that would create significant distortions in international trade and investment. (OECD, 1974).

So the serious question arises: who should bear the cost of maintaining ecological functions and environmental amenities, which are not accounted for in traditional market pricing mechanisms?

Future economic policy will also include a second principle: the 'Beneficiary Compensates' principle. This requires those who benefit from environmental quality to compensate those who maintain the functioning of the ecosystems concerned while using them. The costs compensated for are only the non-market costs. The complexities of applying this principle on a national scale are great, and on an international scale are presently not practical. However, when global benefits are globally recognised, such principles should be internationally accepted as operating for the common good.

The principles of economics and ecology must be integrated in a new economics. As Young (1993) points out, 'Increasingly economists are recognising that ecological systems maintain economic ones and that it is dangerous to assume that economic activity does not influence the capacity of ecological systems to maintain that activity' The inadequacy of present ecological models for predictive purposes is a major constraint to integrating economic and ecological theory and policy.

In the final analysis, the road ahead for Australian development must lie in combining resource-sensitive economic policies established through appropriate institutional arrangements. These policies must strike the right balance of education, incentives and regulations (in that order) to achieve a Sustainable Southland that is economically, ecologically and socially equitable.

GLOSSARY

Acid soil A soil giving an acid reaction throughout most or all of its profile. This kind of soil becomes a problem when its pH drops below 5.5, though any soil with a pH below 7.0 can be considered acidic. At a pH of 5.5 or below, soil is often afflicted by problems such as aluminium, manganese or molybdenum toxicity, which depresses plant and root growth, reducing plant cover and increasing the risk of erosion. Massive application of lime is currently the only way soil acidity is tackled.

Acidification The process whereby soil becomes more acid, which adversely affects plant growth and root nodulation and may cause a decline in plant cover and increase erosion hazard.

Agro-forestry A collective name for land use systems and practices where woody perennials are deliberately integrated with crops and/or animals on the same land management unit. The integration can be either in spatial mixture or temporal sequence. There are normally both ecological and economic interactions between the woody and non-woody components in agro-forestry.

Batter A cut or fill of soil used in road-making, usually with a slope of about 1 in 1.5 and covered with a layer of topsoil to facilitate stabilisation of the roadworks by mat-forming grasses and legumes.

Benthic The deepest stratum of a water body is the benthos, in which the conditions and organisms are termed 'benthic'.

Biodiversity The variety of plant and animal life on earth; encompasses diversity at gene, species and ecosystem levels.

Biomass Matter of biological origin; for example, the living and decaying matter in soil as opposed to the inorganic mineral components.

Catchment An area determined by topographic features within which rainfall will contribute to runoff at a particular point.

Compaction Some soil types, especially clays, can form dense, compacted layers up to several metres below the surface after constant trampling or pressure from tractors and livestock. These layers, or plough pans, can cause waterlogging and other drainage problems which reduce plant growth and increase the likelihood of soil salinity.

Conservation farming A system of using the land within its long-term capabilities. The aim is to conserve the soil, water, fauna and flora of the land, while reaping social and economic benefits from it. Conservation tillage, for instance, aims to reduce the intensity of tillage use and encourage the retention of plant residues — two methods of ensuring the long-term productivity of the land. Types of conservation farming include direct-drilling and minimum or zero tillage.

Contour bank An earth bank constructed to direct water flow over the land to prevent erosion and drainage problems. The bank follows the contour of the land.

Conventional cultivation A tillage system traditionally used for a specific crop in a specific geographical area. Typically includes a sequence of soil workings, such as ploughing, discing and harrowing, to produce a fine seedbed, and the removal of most of the plant residue from the previous crop. In this context the terms 'cultivation' and 'tillage' are synonymous, with emphasis on soil preparation.

Cover crop A crop grown for its value as ground cover to reduce soil erosion, retain soil moisture, provide nitrogen for subsequent crops, control pests, improve soil texture, increase organic matter, or comply with erosion control requirements. Common cover crops include the clovers, vetch, lucerne and rye.

Crop rotation Growing different crops and pastures on the same paddock in annual succession. There are many different types of yearly rotation, i.e. pasture, wheat, legume, wheat. Crop rotation aims to reduce the disease risks of any one crop or pasture and to improve the structure and fertility of the soil.

Cultivation The mechanical preparation of the soil with tractors and implements, usually prior to crop or pasture use. It typically involves a series of soil workings with different types of tools to prepare the soil for seed planting.

Degradation Any decline in the quality of natural resources caused by human activity.

Destocking The removal of stock from a property for an indefinite period.

Die-back A general name for a significant decline in tree numbers and health. It can be caused by several factors, including fungal diseases, insects, pollution and other changes in the natural environment caused by human activity.

Direct drill A reduced tillage system whereby a crop is sown directly into a soil not tilled since the harvest of the previous crop. Stubble and subsequent weed growth are removed by grazing, strategic burning or both. The seedbed may be sprayed with a contact herbicide prior to sowing. Direct drilling is generally practised in mixed farming areas with reliable rainfall, and fallowing is not always necessary.

Ecosystem A community of organisms, interacting with one another, plus the environment in which they live and with which they also interact.

Eutrophication The process by which a body of water becomes rich in nutrients. This can happen naturally or as a result of human activity, e.g. industrial or municipal wastewater or agricultural runoff.

Extension The communication to farmers of practical information and advice; information or technology transfer.

Fallowing Leaving the land uncropped for a period before sowing. Farmers commonly use fallowing to assist weed control and encourage the build-up of moisture and soil nutrients. Fallow length varies from one to twenty months, depending on the climate, the crops grown and the soil type.

Fescue A wintergreen grass of the genus *Festuca,* which is deep-rooted and strongly perennial.

Fertiliser Any substance (manufactured or natural) added to the soil to supplement its nutrient level. The most common fertilisers contain nitrogen, potassium and phosphorus — the three most essential elements for plant growth.

Grazing (or carrying) capacity The maximum sustainable stocking rate of a paddock or property.

Green manure Leguminous crops ploughed into a paddock as an extra source of nitrogen.

Greenhouse effect The effect of gases in the atmosphere that prevent the escape of solar radiation and warm the earth; e.g. carbon dioxide, methane, chlorofluorocarbons (CFCs) and nitrous oxide.

Halophyte A plant able to tolerate and grow in salty (saline) soils.

Herbicide A chemical substance for killing weeds; usually sprayed on crops, pastures or paddocks by tractor-pulled sprayers, and used to replace cultivation in minimum or zero tillage systems.

Hydrology The study of water and its movement over and through soils; important when trying to understand soil salinity, water erosion and waterlogging.

Land capability The land's ability to accept a type and intensity of use permanently, or for specified periods under specific management, without permanent damage. It is an expression of the effect of biophysical land resources, including climate, on the ability of the land to sustain various uses without damage, e.g. crop production requiring regular tillage, grazing, woodland or wildlife.

Land degradation The decline in the quality of the land and its resources, commonly caused by inappropriate human use. It includes soil degradation (or erosion) and the deterioration of natural vegetation, landscapes and water resources, and encompasses the adverse effects of overgrazing, excessive tillage, erosion, sediment deposition, urbanisation, road construction, noxious or feral plants and animals, and decline in native plant communities.

Leaching The removal by water of soluble minerals, nutrients and salts from the soil, usually in a downward direction.

Lime A term encompassing the alkaline calcium materials quicklime (calcium oxide), slaked lime (calcium hydroxide) and limestone (calcium carbonate). Lime most commonly in the form of limestone is applied to soil to neutralise acidity problems or to provide calcium for plant growth.

Macropores The larger spaces between soil aggregates, which are important in increasing the rain infiltration and the moisture holding capacity of soil.

Marginal land Land which, due to its climatic, geographic or topographic location or the nature of its soil and water resources, cannot be used for permanent agriculture.

Medics A group of annual wintergreen legumes of the genus *Medicago,* used for pasture and nitrogen fixation in temperate crop rotations.

Minimum tillage A crop growing system which uses the fewest possible tillage operations to prepare a seedbed. It helps maintain soil structure and soil organic matter. Zero or no-tillage systems rely totally on chemicals, such as herbicides, instead of tillage to control weeds.

Multiple land use Management of an area of land for several purposes at the same time; e.g. agriculture and conservation.

Natural resources Stock of assets established by nature, e.g. mineral resources, soils, air, water, plants and animals.

Nitrogen fixation The chemical transformation of atmospheric nitrogen (N_2) into forms available to plants for growth; accomplished by certain species of symbiotic and free-living bacteria. The more efficient forms are symbiotic with plants, where a food supply and a protected environment are provided to the bacteria in root nodules. The bacteria in turn supply fixed nitrogen to the plant. Strains of the genus *Rhizobium* are the symbiotic, nitrogen-fixing bacteria that associate with leguminous crops such as beans, clover and lucerne. Seeds of leguminous crops are often inoculated with a slurry of *Rhizobium* spores to promote nitrogen fixation by the crop.

No-till A reduced tillage system whereby a crop is sown directly into a soil not tilled since the harvest of the previous crop. All weeds are controlled by herbicides and the subsequent crop is sown directly into undisturbed soil through the remaining stubble and weed residue; all stubble is retained for erosion control. No-till is practised in areas where rain falls mostly in summer and where stock do not predominate. Fallowing is normally required.

Nutrient cycle For an agricultural ecosystem: the input, storage and output of nutrients. For a sustainable system, the inputs must eventually balance the outputs.

Organic Substances or materials derived from plants and animals. Organic farming systems rely upon naturally occurring nutrients, such as those found in cow manure, to replace manufactured fertilisers.

Organic matter Living biota present in the soil or the decaying or decayed remains of animals or plants. The living organic matter in the soil decomposes the dead organic matter. Organic matter in soil increases moisture and soluble nutrient retention, cation (e.g. sodium or calcium) exchange, and water infiltration, and can reduce soil erosion.

Overclearing Any removal of trees, shrubs and other vegetation that makes the land susceptible to degradation.

Overgrazing Continued grazing of a pasture or property at a level that permanently harms its plant, soil and water resources.

Pedogenesis The process of soil formation from bedrock through chemical, physical and biological weathering and soil-forming processes.

Pesticides Chemicals used by farmers to control pests such as weeds (herbicides), insects (insecticides), plant diseases (fungicides), nematodes (nematicides) to regulate plant growth, or to simplify harvest (desiccants).

Piezometry The technique using vertical pipes sunk into the soil to monitor the level of saline groundwater below the soil surface. Used as a measure of the effect of tree planting and other salinity control measures in saline catchments.

Plant nutrient Any element essential to the growth of plants; supplied by fertilisers or occurs naturally in the soil. The major elements are nitrogen, potassium and phosphorus; the minor ones are calcium, magnesium and sulphur; trace elements include iron, molybdenum, manganese and boron.

Point source A term describing pollution which originates from a single site, e.g. factory outfall pipe, as compared to non-point pollution which originates from many diffuse sources, e.g. fertiliser in farm runoff.

Rangeland Land, typically in low rainfall areas, used for extensive sheep and cattle grazing. Rangeland vegetation is native or naturalised pasture.

Reforestation The re-establishment of a forest on a previously cleared or destroyed forest area.

Saline soil A soil containing sufficient soluble salts to affect plant growth and/or land use adversely.

Salinisation The accumulation of free salts in part of a landscape to an extent which causes degradation of vegetation and/or soils. Typically caused by hydrological changes brought about by humans, particularly in areas underlain by marine sediments, i.e. old sea beds.

Sand A soil consisting of particles between 0.02 and 2.00 mm in diameter.

Scald An area of earth made bare by the removal of soil or vegetation by wind or water erosion. Salt scalds are those found on saline land.

Sediment Material of varying size, both mineral and organic, that is being, or has been, moved from its site of origin by the action of wind, water, gravity or ice, and rests on the land either above or below sea level.

Semi-arid A semi-arid region lacks sufficient rainfall to support regular crop production. A semi-arid climate has an average annual rainfall greater than 250 mm and below 375 mm.

Silviculture The production of trees, cf. agriculture (crops), viticulture (vines) and mariculture (fish).

Soil conservation The prevention, mitigation or control of soil erosion and degradation through the application to land of cultural, vegetative, structural and land management measures, either singly or in combination, which enable stability and productivity to be maintained for future generations.

Soil structure The size and shape of soil particles and their arrangement with air spaces into aggregates in the soil profile. The ability of a soil to maintain its aggregated structure under the influence of tillage or rainfall, which tend to disintegrate it, is called the soil structural stability.

Soil structural decline Soil structure is broken down by the aggressive use of tillage equipment, overstocking or excessive cultivation causing denser compact soil. Once the structure of a soil declines, plant growth is repressed and other problems, such as water erosion, become more common.

Soil type A term used to describe a group of soils that can be managed in a similar way or exhibit similar features. There are six major soil types in Australia: sands, cracking clays, calcareous (limey) soils, duplex (two-layered) soils, sesquioxidic (high iron and aluminium) soils and shallow stony soils.

Strip cropping A method of contour planting in conjunction with rotations that alternates strips of crops across the slopes of fields. When practised with conservation tillage, it is a highly effective erosion control method.

Stubble The crop residue that remains after a crop has been harvested.

Stubble incorporation Plant residues partly or wholly buried in the topsoil by mechanical cultivation. Incorporation is more beneficial to soil than burning but much less effective than retaining all residue on the surface.

Sustainable agriculture A system which maintains the productive capacity of the land and its economic viability. It minimises energy and resource use and optimises the rate of turnover and recycling of organic matter and nutrients.

Sustainable use The use of land in a way that does not reduce its future productive capacity.

Systems research Interdisciplinary research that integrates knowledge from several fields of study with research projects designed to generate knowledge and understanding of farming systems.

Tagasaste A drought resistant leguminous fodder tree (known as tree lucerne *Chamaecytisus palmensis*), well adapted to temperate conditions and valuable as a fodder reserve.

Teff An annual hay grass of the genus *Eragrostis*.

Tenure Form of right or title under which property is held.

Tillage The mechanical preparation of the soil by tractor-drawn implements to produce a seedbed for plant growth. See also **Minimum tillage**.

Topsoil The uppermost and usually the most fertile layer of soil.

Water-holding capacity The ability of a soil and crop system to hold water in the root zone.

Waterlogged The term for a soil saturated with water and lacking most or all of the soil air. The condition may be caused by excessive rainfall, poor soil drainage or excessive irrigation.

Wintergreen Frost-resistant pasture species which remain green and nutritious during cold winters.

Wodgil Soil type unique to Western Australia which takes its name from the thin-leafed Acacias which dominate the yellow sandplains.

BIBLIOGRAPHY

Adamson, D.A. and Fox, M.D., 1982. 'Changes in Australasian vegetation since European settlement'. In: *A History of Australasian Vegetation* (ed. J.M.B. Smith). McGraw-Hill, Sydney.

Aitken, D., 1988. 'Countrymindedness — the Spread of an Idea'. In: *Australian Cultural History* (eds. S. Goldberg and I. Smith). Cambridge University Press.

Angus, M., 1975. *The World of Olegas Truchanas*. Australian Conservation Foundation, Melbourne.

Asher, C., 1993. 'Nutrient Depletion — A Long-Term Threat to Grain Production in the North Eastern Region'. Paper to the Queensland Wheat Research Institute, Toowoomba.

Australia and New Zealand Environment and Conservation Council, 1991. *A National Approach to Environmental Impact Assessment in Australia*. ANZECC Secretariat, Canberra.

Australia and New Zealand Environment and Conservation Council, 1992. *Australian Water Quality Guidelines*. National Water Quality Management Strategy, ANZECC Office, Melbourne.

Australian Broadcasting Commission, 1988. *Nature of Australia — a portrait of an island continent*. Part I. Video Bicentennial Series. Festival Records Pty Ltd, Sydney.

Australian Bureau of Agricultural and Resource Economics, 1992. *Agricultural Statistics*. ABARE, Canberra.

Australian Bureau of Agricultural and Resource Economics, 1988–1991. Annual Reports. In: Synnott, M. 'An alternative approach to the institutional, economic and political aspects of sustainable land management'. International Conference on Sustainable Land Management. Nov 1991, Napier, New Zealand.

Australian Bureau of Statistics, 1986. *Population Growth*. Cat. No. 2504.0; 1986. Cat. No. 41150.0. Australian Government Publishing Service, Canberra.

Australian Bureau of Statistics, 1991. *Agricultural Statistics*. ABS, Canberra.

Australian Bureau of Statistics, 1991 & 1992. *Year Book Australia*. ABS, Canberra.

Bakker, R.J., Gordon, I.L. and Robertson, A.G., 1991. 'A survey of organic farming in New Zealand'. Proc. International Sustainable Land Management Conference, Napier, New Zealand.

Barbier, E.B., 1991. 'Environmental Degradation in the Third World'. In: *Blueprint 2 — Greening of the World Economy* (ed. D. Pearce). Earthscan, London.

Barley, K.P., 1959. 'The influence of earthworms on soil fertility'. *Australian Journal of Soil Research* 10: 171–8.

Bates, G., 1992. *Environmental Law in Australia*. Butterworths, Sydney.

Begbie, R., 1989. 'Sustainable Agriculture II: Another Farmer's Viewpoint'. In: *Future* 13: 25–6.

Bennett, H.H., 1939. *Soil Conservation*. McGraw-Hill, New York.

Berry, W., 1977. *The Unsettling of America: Culture and Agriculture*. Sierra Club, San Francisco.

Betts, K., 1993. 'Attitudes to Immigration'. In: *Australians for an Ecologically Sustainable Population. Newsletter* No. 19, Sept: 4.

Birch, C. and Cobb, J.B., 1981. *Liberation of Life*. Cambridge University Press, Melbourne.

Bishop, J.P., 1983. 'Tropical forest sheep on legume forage/fuelwood fallows'. *Agroforestry Systems* 1: 79–84.

Black, A.W., 1994. 'How effective are various policy instruments in promoting the adoption of sustainable farming systems?' In: *Research in Rural Sociology and Development*. Vol. 6 (eds. H. Schwarzweller and T. Lyson). Jai Press, Greenwich, Connecticut.

Black, J.N., 1963. 'The interrelationship of solar radiation and leaf index in determining the rate of dry matter production of swards of subterranean clover'. *Australian Journal of Agricultural Research* 14: 20.

Blaikie, P. and Brookfield, H., 1987. *Land Degradation and Society*. Methuen, New York.

Blyth, M.J. and Kirby, M.G., 1985. 'The Impact of Government Policy on Land Degradation in the Rural Sector'. In: *Policies and Environmental Quality Control* (eds A.J. Jakeman, et al.), CRES, Australian National University, Canberra.

Boer, B.W., 1990. 'Some Legal and Ethical Issues'. In: *Property Rights and Environment Protection* (eds B. Boer and D. James), Environment Institute of Australia.

Bolton, G., 1981. *Spoils and Spoilers: Australians Make Their Environment 1788–1980*. Allen and Unwin, Sydney.

Boulding, K., 1971. 'The Economics of the Coming Spaceship Earth'. In: *Global Ecology* (eds J. Holdren and P. Ehrlich). Harcourt Brace, New York: 180–7.

Boyden, S., 1991. Paper presented to the Fundamental Questions Program, CRES, Australian National University, November.

Bradsen, J., 1988. *Soil Conservation Legislation in Australia*. University of Adelaide. Report to NSCP, Canberra.

Brandt, W., 1980. *North-South: A Programme for Survival*. Pan Books, London.

Bridge, B.J., 1990. 'Organic farming and soil water relations'. In: *Organic Farming in Field Crop Production*, Workshop Series QC91001, QDPI, Toowoomba.

Briggs, J.D. and Leigh, J.H., 1988. *Rare and Threatened Australian Plants*. Special Publication 14, ANPWS, Canberra.

British Medical Association, 1990. *The BMA Guide to Living with Risk*. Penguin Books, London.

Brogen, J.C., 1981. *Nitrogen losses and surface runoff from land spreading of manures*. Nifholff and Junk, The Netherlands.

Brown, L.R., 1990. *State of the World*. Norton, New York.

Brown, R., 1982. 'The Disappearing Wilderness'. *Journal of the Wilderness Society* 18: 7.

Brown, V.A. and Switzer, M.A., 1991. 'Engendering the Debate'. Discussion Paper for Consideration by the Ecologically Sustainable Development Working Groups. Centre for Resource and Environmental Studies, Australian National University, Canberra.

Brown, W.T., 1980. 'Management of Farm Equipment to Minimise Energy Requirements'. In: *Self Sufficiency in Farm Energy*. Agricultural Engineering Society of Australia Seminar Proceedings. Darling Downs Institute of Advanced Education, Toowoomba: 39–42.

Brown, L. and Wolf, E., 1984. Soil Erosion: The Quiet Crisis. Paper No. 60, Worldwatch Institute, Mass., U.S.A.

Bruce, R.C., 1990. 'Soil Nutrients — a free lunch for Australia?' Presidential Address, Australian Society of Soil Science, Queensland Branch, Brisbane.

Brundtland, G.H., 1988. *Our Common Future*. World Commission on Environment and Development, Geneva.

Bryson, R., 1989. 'Environmental opportunities and limits for developments'. Environmental Conservation 16(4): 299–305.

Bureau of Rural Resources, 1991. *Indicators for Sustainable Agriculture* (ed. Ann Hamblin), DPIE, Canberra.

Business Council of Australia., 1990. 'Submission to the Working Party on Ecologically Sustainable Development'. Dept of Environment, Heritage and Sport, Canberra.

Campbell, C.A., 1992. 'Taking the Long View in Tough Times'. Third Annual Report, National Landcare Facilitator, NSCP, Canberra: 10.

Canadian Government, 1992. 'Economic Instruments for Environmental Protection'.

Discussion Paper. Ministry of Supply and Services, Montreal.
Capelin, M.A., 1993. 'Planning Issues Affecting Agricultural Land'. In: *Environmental Soil Science* (eds I.F. Fergus and K.J. Coughlan). Aus. Soc. Soil Sc. Inc., Brisbane.
Capra, F., 1980. *The Turning Point*. Simon and Schuster, New York.
Carson, R., 1962. *Silent Spring*. Houghton Mifflin, Boston.
Carter, V.G. and Dale, T., 1974. *Topsoil and Civilization*. University of Oklahoma Press, Norman, Oklahoma.
Castles, I., 1992. 'Australia's Environment — Issues and Facts'. ABS, Cat. No. 4140.0. AGPS, Canberra.
Caughey, G., Shepherd, N. and Short, J., 1987. Kangaroos: their ecology and management in the sheep rangeland of Australia. Cambridge University Press, Melbourne.
Chambers, R., 1986. *Sustainable Livelihoods*. University of Sussex, Brighton.
Charley, J.L., 1991. 'Effect of post-burn leaching on nutrient status of forest soils'. In: *Fire Research in Rural Queensland* (ed. B.R. Roberts), University of Southern Queensland, 507–17.
Chisholm, A. and Dumsday, R., 1987. *Land Degradation: Problems and Policies*. Cambridge University Press, Melbourne.
Choucri, N., 1991. 'Population and the Global Environment'. In: *Energy and the Environment in the 21st Century* (eds J. Tester, D. Wood and N. Ferrari), MIT Press, Boston: 91–101.
Christie, E.K., 1993. 'International Environmental Conventions and Sustainable Development: A Legal Perspective of the Precautionary Principle'. ASAIHIL Conference Proceedings, Griffith University, Brisbane.
Ciriacy-Wantrup, C.V., 1963. *Resource Conservation*. University of California Press, Berkeley, Calif.
Coastal Zone Inquiry, 1993. Coastal Zone Inquiry Final Report. Resource Assessment Commission. AGPS, Canberra.
Cockfield, G., 1992. 'The Vision Rational: Rural Policy in the 1990s'. Unpubl. Paper. Faculty of Business, University of Southern Queensland, Toowoomba.
Cocks, D., 1992. *Use with Care: Managing Australia's Natural Resources in the Twenty-First Century*. NSW University Press, Sydney.
Coleman, G., 1992. 'The Need for Better Management of Agricultural Land Resources'. Proceedings Australian Agricultural Economics Society. Oct, Brisbane.
Comellack, J.H., 1988. 'Resource allocations for future weed control activities'. 41st NZ Weed and Pest Control Conference. Keynote Address.
Commins, S., 1989. 'Sustainable Development: Sustainable Concept'. *Global Pages*. Jan/March: 2–6.
Commonwealth of Australia, 1991. *Decade of Landcare Plan*. AGPS, Canberra.
Conacher, A. and Conacher, J., 1983. 'A survey of organic farming in Australia'. *Biological Agriculture and Horticulture* 1:241–54.
Conway, G.R., 1987. 'The Properties of Agroecosystems'. *Agricultural Systems* 24: 95–117.
CoResearch, 1981. *CoResearch*, No. 334, CSIRO, Canberra.
Cribb, J., 1990. 'Clean, Green and Clever'. Address to Queensland Landcare Forum, November. QDPI, Brisbane.
Crone, H.D., 1990. *Chemicals and Society — A Guide to the New Chemical Age*. Cambridge University Press, Sydney.
CSIRO, 1989. 'Tree Clearing in the Semi-Arid Tropics'. *Rural Research* 144, Spring: 24–8.
CSIRO, 1990. *A Policy for the Future of Australia's Rangelands*. CSIRO National Rangelands Program, Division of Wildlife and Ecology, Lyneham.
CSIRO, 1991. *Climate Change Scenario*. Climate Impact Group, Division of Atmospheric Research, Canberra.

Daly, H.E., 1980. *Economics, Ecology and Ethics*. Freeman, San Francisco.
Daly, H. and Cobb, J.B., 1989. *For the Common Good: Redirecting the Economic Toward*

Community, the Environment and a Sustainable Future. Beacon Press, Boston.
Daly, J.L., 1989. *The Greenhouse Trap*. Bantam Books, Sydney.
Davidson, B.R., 1987. Appendix B. In: Chisholm and Dumsday, 1987.
De Witt, C.B., 1987. *A Sustainable Earth: Religion and Ecology in the Western Hemisphere*. Only One Earth Forum, May, New York.
Dent, D., 1982. 'The Headaches and Hazards of Farming on the Edge of Urban Areas'. National Farmers Union, *Insight* No. 140, May.
Department of the Arts, Sport, the Environment, Tourism and Territories, 1989. Statistics. DASETT, Canberra.
Department of the Arts, Sport, the Environment, and Territories, 1991. Australian National Report to UNCED, Rio 1992 Conference. AGPS, Canberra.
Department of Home Affairs and Environment, 1981. Australian Environmental Statistics 1980. AGPS, Canberra.
Department of Tourism, 1992. *Tourism: Australia's Passport to Growth — a National Tourism Strategy*, Canberra.
Dorcey, A.H., 1986. *Bargaining in the Governance of Pacific Coastal Resources*. Ch. 3. University of British Columbia, Canada.
Dumsday, R.G. and Edwards, G.W., 1984. 'Economics of Conservation in Agriculture'. In: *Agriculture and conservation achieving a balance* (eds L. Gilmour et al.), Aust. Inst. Agric. Sc., Melbourne.

Ebenreck, Sara, 1983. 'A Partnership Farmland Ethic'. *Canadian Journal of Philosophy*, Vol. 5.
Eckersley, R., 1989. 'Regreening Australia: The environmental, economic and social benefits of reforestation'. CSIRO Occasional Paper No. 3, Canberra.
Economou, N., 1992. 'Problems in Environmental Policy Creation'. In: *Australian Environmental Policy* (ed. K.J. Walker). UNSW Press, Sydney: 41–57.
Edwards, C.A., 1989. 'The importance of integration in sustainable agricultural systems'. *Ecosystems and Environment* 21: 25–35.
Edwards, C.A., 1990. 'The importance of integration in sustainable farming systems'. In: *Sustainable Farming Systems* (ed. C.A. Edwards), Soil and Water Conservation Society, Ankeny, Iowa.
Emery, G.C., 1991. 'Farm Finance — New Challenges'. Queensland Industry Development Corporation. South Qld Rural Outlook Conference, Toowoomba, Oct.
Environment Institute of Australia, 1990. 'Environmental Impact Assessment'. Discussion Paper, EIA, Nathan, Queensland.
Epps, W.R. and Crittenden, R., 1992. 'Appraisal of Land Degradation in Australia: rectification of agricultural land management'. *Land Use Policy* 9: 199–208.
ESD, 1991. Intersectoral Issues Report of the Ecologically Sustainable Development Working Group. AGPS, Canberra.

Fitzgerald, R., 1988. 'Immigration: A Commitment to Australia'. Report of Committee to Advise on Australia's Immigration Policies. AGPS, Canberra.
Fletcher, M. and Southwell, C.J., 1990. *Kangaroo Population Trends in the Australian Rangelands*. ANPWS, Canberra.
Freudenberger, D., 1993. In: T. Johnstone, 'Farm kangaroos, shoot your sheep'. *Australian Farm Journal*, Oct: 32–3.
Friend, J.A., 1992. 'Achieving Soil Sustainability'. *Journal of Soil and Water Conservation* 47(2): 156–7.
Gaffney, J., 1990. 'Organic farming economics and market prospects'. In: *Organic Farming in Field Crop Production*, Workshop Series QC91001, QDPI, Toowoomba.
Gardner, A., 1989. 'A Consensus System of Planning and Management for Land Conservation'. *Environmental and Planning Law Journal* 6: 249–70.
Georgescue-Roegen, N., 1980. 'Entropy Law and the Economic Problem'. In: *Economics, Ecology and Ethics* (ed. H. Daly). W. Freeman, San Francisco: 49–81.

Gifford, R.M., Kalma, J.D., Aston, A.R. and Millington, R.J., 1975. 'Biophysical Constraints in Australian Food Production: Implications for population policy'. *Search* 6(6): 212–13.

Gillard, E., Brough, E. and Duffield, B., 1990. 'Extension Strategies Involving Rural Women'. Qld Dept Primary Industries, Brisbane, November.

Goucher, G., 1994. 'New Income Averaging Plan'. *Queensland Country Life,* 3 Feb, Brisbane, p. 11.

Gow, L., 1989. 'Role of Indigenous Peoples in Environmental Assessment'. *Environmental Update* No. 115, Wellington, New Zealand.

Grace, P. and Oades, J.M., 1994. 'Long-term Field Trials in Australia'. Unpubl. ms. Waite Institute, Adelaide.

Graetz, R.D., Walker, B.H. and Walker, P.A., 1988. 'The consequences of climatic change in seventy percent of Australia'. In: *Greenhouse, planning for climatic change* (ed. G.I. Pearman). CSIRO, Canberra.

Grierson, I., 1985. USLE Teaching Unit. Agricultural College, Roseworthy, South Australia.

Griffiths, T., 1991. 'History and Natural History'. In: *Packaging the Past* (eds J. Richard and P. Spearritt). Melbourne University Press: 16–32.

Grigg, G., 1988. 'Conservation of the Sheep Rangelands — Are kangaroos a better economic base?' *Acres Australia* 1: 17–19.

Grigg, G., 1993. 'Are we farming the wrong animals?' *Australians for an Ecologically Sustainable Population Newsletter* No 18: 1–4.

Hallsworth, E.G., 1977. 'A Forward Look at Land Use Policy for Australia'. *Search* 8(3): 58–67.

Hamblin, A., 1991. 'Sustainability: Physical and Biological Considerations for Australian Environments'. Bureau of Rural Resources Working Paper 19/89, rev. edn.

Hamblin, A., 1992. 'Environmental indicators for sustainable agriculture'. Report on a national workshop, Nov 28–9, 1991. Bureau of Rural Resources, Land and Water Resource Research and Devt Corp, Grains Research and Devt Corp, Canberra.

Hamblin, A. and Goss, K., 1993. 'Sustainable Agriculture Indicators for Australia and New Zealand'. SCARM Report No. 51. Dept Primary Industries and Energy, Canberra.

Hamblin, A. and Kyneur, G., 1993. *Report on Australian Wheat Yields 1950–1991.* National Resource Information Centre, Bureau of Resource Sciences, Canberra.

Hart, R.D., 1986. 'Ecological framework for multiple cropping research'. In: *Multiple Cropping Systems* (ed. C.A. Francis), Macmillan, New York.

Hassall and Associates, 1990. 'The Market for Australian Produced Organic Food'. Report for Special Rural Research Council, Canberra.

Heathcote, R.L., 1965. *Back of Bourke: A Study of Land Appraisal and Settlement in Semi-Arid Australia.* Melbourne University Press.

Henderson-Sellers, A. and Blong, R., 1989. *The Greenhouse Effect: Living in a Warmer Australia.* NSW University Press, Sydney.

Hendrix, P.F., Crossley, D.A., Blair, J.M. and Coleman, D.C., 1990. 'Soil biota as components of sustainable agroecosystems'. In: *Sustainable Agricultural Systems* (eds C.A. Edwards, R. Lal, P. Madden, R.H. Miller and G. House). Soil and Water Conservation Society, Ankeny, Iowa.

Henschke, C.J., 1980. 'Saltland in Statistics, the 1979 Saltland Survey'. *Journal of Agriculture of WA* 1(4).

Herberern, J., 1992. 'A Soil Health Index'. *Journal of Soil and Water Conservation* 47(1): 1992.

Herrmann, T., 1993. 'Crop Rotation Sustainability Index'. Soil and Water Conservation, South Aust Dept Primary Industries.

Highsmith, R.M., 1969. *Conservation in the United States.* Rand McNally, Chicago.

Holden, A., 1992. *The Cape York Peninsula Land Use Strategy (CYPLUS).* Holden and Associates, Brisbane.

Holmes, J., 1990. 'Why Persist with Pastoral Leasehold Tenures?' Arid Land Administrators Conf, Dept Lands, Charleville: 25–31.

Honnef, M., 1986. 'Analysis of an organic farming system that integrates livestock and grain production on the Darling Downs'. BSc(Agric) Thesis, Queensland Agricultural College, Gatton.

Hudson, N., 1976. *Soil Conservation*. Batsford Press, London.

Hudson, N., 1992. 'From Soil Conservation to Land Husbandry'. *Australian Journal of Soil and Water Conservation* 5(2): 4–8.

Intergovernmental Agreement on the Environment, 1992. Commonwealth of Australia, Canberra.

International Federation of Agricultural Producers, 1991. 'Sustainable Farming and the Role of Farmers' Organisations'. *World Farmers' Times* 1/91.

Jacks, G.V. and Whyte, R.O., 1939. *Rape of the Earth — A World Survey of Soil Erosion*. Faber, London.

Jackson, W., 1984. 'Toward a unifying concept for agriculture'. In: *Agricultural Sustainability in a Changing World Order* (ed. G. Douglass), Westview Press, Boulder, Col.: 159–74.

James, D., 1993. 'Australia's experience in using economic instruments for meeting environmental objectives'. Discussion Paper, DEST, Canberra.

Jenny, H., 1941. *Factors of Soil Formation*. McGraw-Hill, New York.

Kahn, H., Brown, W. and Martel, L., 1976. *The Next 200 Years: A Scenario for America and the World*. Morrow & Co. Inc., New York, p. 10.

Keys, M.G., 1984. 'Stand changes, tree damage and site disturbance caused by selective logging in rainforests'. In: *Managing the Tropical Rainforest* (eds K. Shepherd and H. Richter), Australian National University, Canberra.

Lines, W.J., 1991. *Taming the Great South Land: a history of the conquest of Nature in Australia*. Allen and Unwin, Sydney.

Littleboy, M., Silburn, D.M., Freebairn, D.M., Woodruff, D.R. and Hammer, G.L., 1989. 'Wheat yield reductions due to predicted soil erosion under three fallow managements in the Central Highlands'. In: *Perfect — A computer simulation model of productivity erosion runoff functions to evaluate conservation techniques*. Dept Primary Industries, Queensland Govt Bulletin QB89005.

Lodge, G.M., 1987. 'An outline of the current program for the domestication of *Danthonia linkii*'. Domestication of Native Grasses Workshop, Tamworth. Dept. Agric., NSW: 10–11.

Lodge, G.M. and Groves, R.H., 1990. 'The Domestication and Agronomy of Native Grasses'. Native Grass Workshop. Australian Wool Corporation, Oct, Dubbo.

Lovelock, J.E., 1989. *Gaia: A New Look at Life on Earth*. Oxford University Press, Oxford.

Lynch, J.M., 1987. 'Soil Biology: Accomplishments and Potential'. *Soil Science Society of America Journal* 51: 1409–12.

McConnell, K.E., 1983. 'An Economic Model of Soil Conservation'. *American Journal of Agricultural Economics* 65(1): 83–89.

McCown, R.L., 1991. 'Research in a Farming Systems Framework'. In: *Dryland Farming* (eds V. Squires and P. Tow), Oxford University Press, Melbourne: 241–9.

McElroy, M.B., 1985. 'Change in the Natural Environment of the Earth'. In: *Sustainable Development of the Biosphere* (eds C. Clerk and R. Nunn), International Institute of Applied Systems Analysis, Laxenburg, Austria: 199–212.

McHarg, I.L., 1969. *Design with Nature*. Natural History Press, New York.

McHarg, I.L., 1992. 'Green the Earth, Heal the Earth'. *Journal of Soil and Water Conservation* 47(1): 39–41.

Mackay, N., 1992. 'A Nutrient Management Strategy for the Murray-Darling Basin'. In:

Catchments of Green Conference Proceedings, Greening Australia, Adelaide: 83–90.

McKenzie, N.J., 1991. 'A Strategy for Coordinating Soil Survey and Land Evaluation in Australia'. Division of Soils Report No. 114, CSIRO, Canberra.

McKeon, G.M. and Howden, S.M., 1988. 'The effect of climatic change on crop and pastoral production in Queensland'. In: *Greenhouse: Planning for Climatic Change* (ed. G. Pearman). CSIRO, Brill, New York.

Meadows, D., Meadows, D. and Randers, J., 1992. *Beyond the Limits*. Chelsea Green Publishing Co., London.

Messer, J., 1987. 'The Sociology and Politics of Land Degradation in Australia'. In: *Land Degradation and Society* (eds P. Blaikie and H. Brookfield), Methuen, New York.

Mills, D., 1991. 'Women in Landcare — Nurturing an Ethic'. In: *Runs on the Board*. Annual Conf. Soil and Water Conservation Assoc. of Australia, Toowoomba, Nov.

Mills, J.R., Turner, E.J. and Catabiano, T., 1992. *Land Degradation in South-Western Queensland*. Queensland Department of Primary Industries, Brisbane.

Moore, E., 1990. 'A Sustainable Population for Australia: Dilemma for the Green Movement'. Masters Thesis, Mawson Graduate Centre, University of Adelaide.

Morrison, D., Bathgate, A. and Barton, J., 1991. 'Modelling to Analyse an Integrated Crop-Livestock Farm — MIDAS'. In: *Dryland Farming* (eds V. Squires and P. Tow), Oxford University Press, Melbourne.

Morton, S.R. et al., 1994. 'The Stewardship of Arid Australia: Ecology and Landscape Management'. *Journal of Environmental Management* (in press).

Moss, A.J. et al., 1993. 'A Preliminary Assessment of Sediment and Nutrient Exports from Queensland Coastal Catchments'. Environment Technical Report No. 5, QDPI, Brisbane.

Mott, J.J., Williams, J., Andrew, M.H. and Gillison, A.N., 1985. 'Australian Savanna Ecosystems'. In: *Ecology and Management of the World's Savannas* (eds. J. Tothill and J. Mott), Australian Academy of Science, Canberra.

Nash, R., 1982. *Wilderness and the American Mind*. New Haven, Connecticut.

Nash, R., 1991. *The Rights of Nature*. Primavera Press, Leichhardt, New South Wales.

National Research Council, 1986. *Assessing the National Resources Inventory*. National Academy Press, Washington DC.

National Research Council, 1989. *Alternative Agriculture*. National Academy Press, Washington DC.

National Research Council, 1991. 'Toward Sustainability: a plan for collaborative research on agriculture and natural resource management'. Board on Agriculture Report. National Academy Press, Washington, DC.

Nelson, P., 1990. 'Agricultural Systems for the Future'. Wheatbelt Landscape Conference, Tammin, W.A.

New South Wales Planning and Environment Commission, 1979. 'Interim Report on Viability of Rural Activities'. June, Sydney.

Nicholson, D.I., 1984. 'The Development of Silviculture in North Queensland Rainforests'. In: *Managing the Tropical Forest* (eds K. Shepherd and H. Richter), Australian National University: 203–19.

Nowak, P., 1992. 'The Value of Values in Resource Management'. *Journal of Soil and Water Conservation* 47(5): 356–9.

Oades, J.M. and Walters, L.J., 1994. 'Indicators for Sustainable Agriculture: Policies to Paddocks'. International Workshop on Soil Biota: Management in Sustainable Farming Systems. CSIRO, Adelaide.

Oakley, G., 1991. 'The Statistics of Resource and Environmental Appraisal'. In: *Environmental Indicators for Sustainable Agriculture*. Report on a National Workshop (ed. Ann Hamblin), Bureau of Rural Resources, DPIE, Canberra.

Odum, E. and Franz, E., 1980. 'Whither the life support system?' In: *Growth without Ecodisasters* (ed. N. Polunin), Halsted-Wiley, New York.

OECD, 1974. *The Polluter Pays Principle: definition, analysis and implementation.* OECD Office, Paris.

OECD, 1991. *Environmental Policy: How to Apply Economic Instruments.* OECD Office, Paris.

Office of Local Government, 1992. *Principles for approvals reform at local government level.* OLG, Canberra.

Office for the Status of Women., 1991. Report to 'Partners in Life' — Global Assembly on Women and the Environment. Miami, Florida, Nov.

O'Riordan, T., 1971. *Perspectives on Resource Management.* Pion Ltd, London.

O'Riordan, T., 1988. 'The Politics of Sustainability'. In: *Sustainable Environmental Management* (ed. R.K. Turner). Belhaven Press, London.

Orr, D.M., 1981. 'What has this drought done to pasture stability?' *The Mulga Line.* QDPI, Charleville, 10/81: 6.

Osborne, F., 1963. *Our Crowded Planet.* Allen and Unwin, London.

Papendick, R., 1987. 'Why consider alternative production systems?' *American Journal of Alternative Agriculture* 2(2): 83–6.

Pearce, D., Markandya, A. and Barbier, E.B., 1989. *Blueprint for Green Economy.* Earthscan Publications, London.

Pearce, H. and Bradley, K., 1983. 'The Threat to the Environment of the WA Land Release Program'. In: *What Future for Australia's Arid Lands?* (eds. J. Messer and G. Mosley), Australian Conservation Foundation, Melbourne.

Pickup, A. and Stafford-Smith, D.M., 1993. 'Problems, Prospects and Procedures for assessing the sustainability of pastoral land management in arid Australia'. *Global Ecology and Biogeography Letters*, in press.

Pittock, A.B., 1989. 'Environmental Changes and their effects on Natural Hazards'. Proceedings of the Sterling Reinsurance Natural Disasters Seminar 1/9/89, Sydney.

Poincelot, R.P., 1987. *Toward a more sustainable agriculture.* AVI Publishing Co. Inc., Westport, Connecticut.

Pressland, A.J. and Anderson, E.R., 1992. Draft Manual for QGRAZ: Monitoring the Condition of Grazing Lands. QDPI, Brisbane.

Proofs, J.L., 1989. 'Ecological Economics'. *Ecological Economics* 1(1): 59–76.

Queensland Department of Primary Industries, 1990. 'Organic Farming in Field Crop Production'. Conference and Workshop Series QC91001, Dept Primary Industries, Brisbane.

Rawls, J., 1971. *A Theory of Justice.* Horwood University Press, Cambridge.

Reeve, I. and Black, A.W., 1993. 'Australian Farmers' Attitudes to Rural Environmental Issues'. Report, Rural Development Centre, University of New England, Armidale.

Richardson, G., 1994. 'Whole Farm Planning — Becoming a Way of Life in Tasmania'. *Australian Journal of Soil and Water Conservation* 7(1): 9–18.

Rickert, K.G. and McKeon, G.M., 1991. 'Modelling as an Aid to Systems Evaluation'. In: *Dryland Farming — A Systems Approach* (eds V. Squires and P. Tow). Sydney University Press: 208–21.

Roberts, B.R., 1960. 'Studies on irrigated pastures'. MSc(Agric) Thesis, University of Natal.

Roberts, B.R., 1965. 'Applied plant ecology in land use planning of catchments'. *S. Afr. J. Science*, 61: 111–17.

Roberts, B.R., 1971. 'The Role of Animals in Maintaining Soil Fertility'. *Fertilizer Society of South Africa Journal* 2: 17–22.

Roberts, B.R., 1984a. 'Land Ethics — a necessary addition to Australian Values'. In: *Birth of Land Care.* University of Southern Queensland Press, Toowoomba.

Roberts, B.R., 1984b. *Soil Conservation.* Study Book for Continuing Education. Darling Downs Institute of Advanced Education, Toowoomba, Queensland.

Roberts, B.R., 1987. 'Australian Land Degradation and its Control'. *AMBIO, a Journal of the Human Environment.* Royal Swedish Academy of Sciences/ Pergamon Press, Vol. 16, No. 5: 272–6.

Roberts, B.R., 1990a. 'Potential of Native Grasses and Barriers to Maximising their Use'. In: *Native Grass Workshop Proceedings*, Australian Wool Corporation, Dubbo, October.

Roberts, B.R., 1990b. 'The Big Picture: Suggested Planning Framework for Queensland Landcare 1990–2000'. Unpubl. Discussion Document for State Landcare Council, Brisbane, March.

Roberts, B.R., 1991. 'Adoption of the Land Ethic in Australia: Its Origin, Development and Monitoring'. In: *Runs on the Board.* Annual Conf. Soil and Water Conservation Assn of Australia, Toowoomba, Nov.

Roberts, B.R., 1992a. *Land Care Manual.* University of New South Wales Press, Sydney: 51.

Roberts, B.R., 1992b. 'Sustainable Animal Production — The Role of Producer Organisations'. Cattlemen's Union Conf, Toowoomba.

Roberts, B.R., 1992c. 'Sustainable Land Use — What can a Christian Land Ethic Contribute?' In: *Ground Rules — Perspectives on Land Stewardship.* University of Southern Queensland Press, Toowoomba: 267–80.

Roberts, B.R., 1992d. 'Landcare at Rio'. *Landcare News* No. 6. QDPI, Brisbane.

Roberts, B.R., 1993. *Ground Rules — Perspectives on Land Stewardship.* University of Southern Queensland Press, Toowoomba.

Roberts, B.R. and Scott, J.D., 1968. 'The Nutritive Value, Yield and Botanical Composition of Irrigated Pastures as influenced by nitrogen fertilizing'. *Grassland Society of Southern Africa* 3: 83–9.

Roberts, B.R. and Silcock, R.G., 1993. *Western Grasses* (2nd edn). University of Southern Queensland Press, Toowoomba.

Robertson, G.A., 1983. 'The Cost of Neglecting Conservation Practices'. In: *Sand, Salt and Farming — A Land Use Review.* Dept Agric Perth, 17 March.

Robertson, G.A., 1990. 'The Wheatbelt environment in the year 2000'. In: Wheatbelt Landscape Conference audiotape set. Tammin Landcare Group, Tammin, WA.

Robertson, L.N., 1990. 'Insect pest control in field crops without synthetic chemicals'. In: *Organic Farming in Field Crop Production*, Workshop Series QC91001, QDPI, Brisbane.

Russell, J.S., 1973. 'Yield trends in different crops in different areas and reflections on the sources of crop yield improvement in the Australian environment'. *Journal of the Australian Institute of Agricultural Science* 39: 156–166.

Salter, R., 1989. 'Sustainable Futures: Global Inequality fuels the Environmental Crisis'. *In Future.* Issue 13, July, AFCOA, Canberra: 14–16.

Sears, P.B., 1959. 'The Ecology of Man'. Annual Report, Smithsonian Institute, Washington D.C.

Seattle, Chief., 1854. In: *Indian Oratory* (ed. W. Vanderworth). Norman Publications, Oklahoma, 1971.

Silburn, D.M., Carroll, C., Ciesiolka, C.A. and Hairsine, P., 1992. 'Management Effects on Runoff and Soil Loss from Native Pasture in Central Queensland'. 7th Australian Rangeland Society Conf. Working Papers, Cobar: 294–5.

Silvey, M.W., Simpson, B.W. and Silvey, P.J., 1993. 'Environmental Chemistry of Organic Chemicals and Biocides'. In: *Environmental Soil Science* (eds. I. Fergus and K. Coughlan), Aus. Soc. Soil Sc., Brisbane.

Solow, R.M., 1974. 'Intergenerational equity and exhaustible resources'. Summed up in *Scandinavian Journal of Economics* 88: 141–149.

Squires, V., 1991. 'A Systems Approach to Agriculture'. In: *Dryland Farming* (eds V. Squires and P. Tow), Oxford University Press, Melbourne: 4–15.

Stadler, T., 1989. 'Agricultural Land Degradation in Southern Australia'. PhD Thesis, U. Tasmania, Hobart.

Stafford-Smith, M., 1994. 'Sustainable Production Systems and Natural Resource Management in the Rangelands'. In: *Outlook 94*, Australian Agricultural Outlook Conf, ABARE, Canberra: 148–59.

Standing Committee on Agriculture, 1991. Ecologically Sustainable Development Working Groups. Final Report — Agriculture. Australian Government Publishing Service, Canberra.

Stinner, B.R. and Blair, J.M., 1990. 'Ecological and agronomic characteristics of innovative cropping systems'. In: *Sustainable Agricultural Systems* (eds C.A. Edwards et al.), Soil and Water Conservation Society, Ankeny, Iowa.

Talbot, F., 1976. 'Environmental Impact Assessment: Summary and Prospects'. Search 7(6): 273–4.

Teese, C., 1989. 'Immigration and Growth — Myth and Reality'. Paper to Australians for an Ecologically Sustainable Population Conf, ANU, 7 Oct, Canberra.

Thompson, J.P., 1990a. 'How does organic farming perform in relation to soil biology?' In: *Organic Farming in Field Crop Production*, Workshop Series QC91001, QDPI, Toowoomba.

Thompson, P.J.M., 1990b. 'Role of livestock in organic farming of field crops'. In: *Organic Farming in Field Crop Production.*

Tisdell, C., 1990. 'The Nature of Sustainability and of Sustainable Development'. Discussion Paper No. 48, Dept Economics, University of Queensland, Brisbane.

Tothill, J.C., Nix, H.A., Stanton, J.P. and Russell, M.J., 1985. 'Land Use and Productive Potentials of Australian Savanna Lands'. In: *Ecology and Management of the World's Savannas* (eds J.C. Tothill and J.J. Mott), Australian Academy of Science, Canberra.

Twyford, J., 1980. 'Oil crops for the Darling Downs and Queensland'. In: *Self-sufficiency in Farm Energy*. Darling Downs Institute of Advanced Education, Toowoomba: 32.

UN, 1972. Report on the Stockholm Conference on the Human Environment. United Nations Organisation, New York.

UNCED, 1992. United Nations Conf, Environment and Development. Rio de Janeiro, Brazil.

United Nations Environment Program, 1988. 'Environmental Impact Assessment. Basic Procedures for Developing Countries'. Regional Office for Asia and the Pacific, UNO, New York.

University of California, 1990. 'Sustainable Agriculture: Balancing Social, Environmental and Economic Concerns'. Conf. Proceedings, Santa Cruz, June.

USDA, 1987. Agricultural Resources — Cropland, Water and Conservation. Report AR-8, Economic Research Service, Washington, DC.

Vance, P.N., 1990. 'How does organic farming perform in relation to whole farm management?' In: *Organic Farming in Field Crop Production*, Workshop Series QC91001, QDPI, Toowoomba.

Wadham, S.M. and Wood, G.L., 1939. *Land Utilization in Australia.* Melbourne University Press.

Walker, J., Hatton, T., and Mundy, D., 1992. 'The CSIRO Land and Water Care Program'. In: *Catchments of Green Conference Proceedings*, Greening Australia, Adelaide: 7–15.

Walther, P., 1986. 'The Meaning of Zoning in the Management of Natural Resource Lands'. *Journal of Environmental Management* 22: 331–43.

Westbury, V., 1991. 'Ecofeminising Australia'. *The Bulletin*, 2 July: 89–91.

Westman, W., 1987. *Ecology, Impact Assessment and Environmental Planning.* Wiley, New York.

Wheeler, J.L. and Freer, M., 1973. 'Pasture and Forage: The Feed Base for Pastoral Industries'. In: *The Pastoral Industries of Australia* (eds G. Alexander and G. Williams), Sydney University Press: 165–82.

White, I.D., Mottershead, D.N. and Harrison, S.J., 1984. *Environmental Systems — An*

Introductory Text. Allen and Unwin, Sydney.

Williams, J., 1989. 'Land Degradation: Evidence that current Australian farming practice is not sustainable'. In: *Management for Sustainable Farming* (ed. R.J. Hampson), Australian Farm Management Society Conf. Proc., Emerald: 165–190.

Williams, J. and Hamblin, Ann, 1991. 'Regions of Australia: Issues for Sustainable Agriculture'. Unpubl. Report to Working Party on Sustainable Development in Agriculture. AGPS, Canberra.

Williams, O.B., 1964. 'Energy Flow and Nutrient Cycling in Ecosystems'. *Australian Society of Animal Production* V: 291–300.

Wittwer, S.H., 1975. 'Food Production — technology and the resource base'. *Science* 188: 597–8.

Working Group on Sustainable Agriculture, 1991. 'Sustainable Agriculture'. Report of the Working Group on Sustainable Agriculture. SCA Technical Report Series No. 36, AGPS, Canberra.

World Commission, 1978. Report on Environment and Development. United Nations Environment Program, Geneva.

World Council of Churches, (1974). 'Science and Technology for Human Development'. *Anticipation* 19: 1–43, Bucharest.

World Resources Institute, 1985. *World Resources 1985*. Oxford University Press, New York.

World Resources Institute, 1986. *World Resources 1986*. Oxford University Press, New York.

Wynen, E. and Edwards, G., 1990. 'Towards a comparison of chemical-free and conventional farming in Australia'. *Australian Journal of Agricultural Economics* 34(1): 39–55.

Wynne, R., 1992. 'Economics of Stocking Rates in the Semi-Arid Rangelands of NSW'. 7th Australian Rangelands Society Conf, Working Papers, Cobar: 378–9.

Yapp, T., 1989. 'The Costs of Degradation to the Community: Issues and Estimates'. *Australian Journal of Soil and Water Conservation* 2(3): 32–6.

Young, C., 1989. 'Understanding Population Growth in Australia'. Paper, Australians for an Ecologically Sustainable Population Conf, ANU, 7 Oct, Canberra.

Young, M.D., 1985. 'Pastoral Land Tenure Options for Australia'. *Australian Rangeland Journal* 7(1): 43–6.

Young, M.D., 1990. 'Arid Land Administration and Tenure: The 1980s and 1990s'. Arid Land Administrators Conf. Dept Lands, Charleville: 30–40.

Young, M.D., 1992a. *Sustainable Investment and Resource Use: Environmental Integrity and Economic Efficiency*. UNESCO, Paris.

Young, M.D., 1992b. *Sustainable Investment and Resource Use: Man and the Biosphere Series*. Vol. 9. Parthenon, New Jersey.

Young, M.D., 1993. 'For our Children's Children: Some Practical Implications of Inter-Generational Equity and the Precautionary Principle'. *Resource Assessment Commission, Occasional Publication* No. 6, Nov.

Zimmerman, E.W., 1951. *World Resources and Industries*. Harper and Row, New York.

SUGGESTED FURTHER READING

Allen, Patricia, *Food for the Future: Conditions and Contradictions of Sustainability*, Wiley Books, 1993.

Chisholm, A. and Dumsday, R. (eds), *Land Degradation: Problems and Policies*, Cambridge University Press, 1987.

Cocks, Doug, *Use With Care: Managing Australia's Natural Resources in the Twenty-First Century*, NSW University Press, 1992.

Corey, S., Dall, D., and Milne, W. (eds), *Pest Control in Sustainable Agriculture,* CSIRO, Canberra, 1993.

Edwards, C., Lal, R., Madden, P., Miller, R., and House, G. (eds), *Sustainable Agricultural Systems* (4th Printing), St Lucie Press, 1993.

Edwards, C., Wall, M., Horn, D., and Miller, F. (eds), *Agriculture and the Environment*, Elsevier, 1993.

Greenland, D., and Szaboics, I. (eds), *Soil Resilience and Sustainable Land Use*, CAB, 1994.

Hatfield, J. and Karlen, D. (eds), *Sustainable Agriculture Systems*, Lewis Publishers, Iowa, 1993.

Kula, E., *Economics of Natural Resources, the Environment and Policies*, Chapman Hall, 1994.

Lutz, Ernst (ed.), *Toward Improved Accounting for the Environment*, World Bank, 1993.

Mercer, David, *A Question of Balance: Natural Resources Conflict Issues in Australia,* Federation Press, 1991.

Mulongoy, K., Gueye, M., and Spencer, D. (eds), *Soil Organic Matter Dynamics and the Sustainability of Tropical Agriculture*, Wiley, 1993.

National Research Council, *Alternative Agriculture*, National Academy Press, Washington DC, 1989.

National Research Council, *Sustainable Agriculture and the Environment in the Humid Tropics*, National Academy Press, Washington, DC 1993.

National Research Council, *Rangeland Health*, National Academy Press, Washington DC, 1994.

OECD Publications, *Environment and Taxation,* 1994.

Roberts, B., *The Land Care Manual,* NSW University Press, 1992.

Roberts, B., *Ground Rules: Perspectives on Land Stewardship,* USQ Press, 1993.

Russell, J., and Isbell, I. (eds), *Australian Soils — The Human Impact,* UQ Press, 1986.

Smith, David, *Saving a Continent: towards a sustainable future*, University of New South Wales Press, 1994.

Squires, V., and Tow, P., *Dryland Farming: A Systems Approach,* Oxford University Press, 1991.

Tan, Kim, *Environmental Soil Science*, Dekker, 1994.

Turner, Kerry (ed.), *Sustainable Environmental Economics and Management*, Pinter, 1992.

van Lier, H., Jaarsma, C., and Jurgens, C. (eds), *Sustainable Land Use Planning*, Elsevier, 1994.

Widdowson, R., *Towards Holistic Agriculture*, Permagon, 1987.

Wiebers, Uwe-Carsten, *Integrated Pest Management and Pesticide Regulation in Developing Asia,* World Bank Publications, 1993.

INDEX